Contemporary Topics in

POLYMER SCIENCE

Volume 3

Contemporary Topics in

POLYMER SCIENCE

SCIENCE

Volume 3

Edited by
Mitchel Shen
University of California
Berkeley, California

SPRINGER SCIENCE+BUSINESS MEDIA, LLC

Library of Congress Catalog Card Number 79-641014

ISBN 978-1-4615-6742-4   ISBN 978-1-4615-6740-0 (eBook)
DOI 10.1007/978-1-4615-6740-0

Proceedings of the Ninth Biennial Polymer Symposium of the
Division of Polymer Chemistry held at Key Biscayne,
Florida on November 18—22, 1978

© 1979 Springer Science+Business Media New York
Originally published by Plenum Press, New York in 1979
Softcover reprint of the hardcover 1st edition 1979

Mitchel Shen died from an undiagnosed illness on August 7, 1979. He edited this volume and served as the organizer and chairman of the Ninth Biennial Symposium of the Division of Polymer Chemistry of the American Chemical Society. The following tribute to this outstanding polymer scientist was prepared at my request by a close associate, Professor Adi Eisenberg of McGill University.

E. J. Vandenberg
Chairman, Division of
Polymer Chemistry, Inc.
August 30, 1979

Mitchel Shen

Mitchel Shen
1938 - 1979
In Memoriam

Mitchel Shen was born in Teinjin, China, September 1, 1938 and came to the United States in 1955. He received a B.S. degree at St. Francis College, Pennsylvania, and pursued his graduate education at Princeton with A. V. Tobolsky. He joined the staff of the North American Rockwell Science Center in Thousand Oaks, California in 1964 and in 1969 he moved to the University of California, Berkeley, where he was a Professor of Chemical Engineering until his untimely death.

His entire professional career was devoted to research and teaching in the field of high polymers. His wide range of interests and high productivity have resulted in contributions to fields as diverse as plasma polymerization, rubber elasticity, the glass transition, viscoelastic properties, and many other areas spanning the range from polymer synthesis to theoretical polymer physics. In addition to well over 100 publications, he wrote or edited 7 books. His managerial talents were evident in his capacity as vice-chairman of the department, treasurer of the Division of Polymer Chemistry, Co-Editor of the Reviews in Macromolecular Science and organizer of various symposia, three of them during the past year alone.

Mitchel Shen's premature passing represents a tragic loss not only to his immediate family, his wife, Vivian, two young children, Teddy and Lara, his parents, brothers and sisters, but also to the scientific community at large where he clearly belonged among the most brilliant and creative. He had achieved a lot for one so young, and the future held promise of much more to come. He will be sorely missed by his many students and co-workers, and by all those with whom he came into contact during his all too short productive career.

Adi Eisenberg

Professor Carl Shipp Marvel
Winner of the 1978
Division of Polymer Chemistry Award
American Chemical Society

PREFACE

The Ninth Biennial Polymer Symposium, sponsored by the Division of Polymer Chemistry of the American Chemical Society, was convened during November 18-22, 1978 at Key Biscayne, Florida.

The symposium also marks the occasion when the Second Division of Polymer Chemistry Award was presented to Professor C. S. Marvel for his outstanding achievements in polymer chemistry. The proceedings of this Symposium are recorded in this volume.

Carl Shipp Marvel, more popularly known as Speed Marvel, was born in Waynesville, Illinois in 1894. He received his A.B. and M.S. degrees at Illinois Wesleyan University, and M.A. and Ph.D. degrees at the University of Illinois. After having completed his graduate education, Professor Marvel remained at the University as a member of the Chemistry faculty. He "retired" from Illinois in 1961, and moved to the University of Arizona where he continued his research in polymer chemistry. Out of the more than 500 publications, over 100 were from Arizona during his active retirement. Speed has so far trained 176 Ph.D. students and 128 postdoctoral fellows, and inspired countless young chemists to the science of macromolecules. No doubt these numbers will increase along with his scientific productivity.

Professor Marvel has been honored on numerous ocasions prior to receiving the Division of Polymer Chemistry Award. He is the recipient of the Nichols Medal, Willard Gibbs Medal, American Institute of Chemists Gold Medal, Priestley Medal, SPE International Award, Witco Award, Perkin Medal, Madison Marshall Award, American Institute of Chemists Pioneer Award, John Kuebler Award, and Borden Award. He is also a member of the National Academy of Sciences as well as the National Academy of Arts and Sciences. Speed has received honorary degrees from the Illinois Wesleyan University, University of Illinois, and the University of Louvain.

   This volume begins with the proceedings of the Award
Ceremoney for Professor Marvel. His Award Address was
followed by a panel discussion in which the human side of
Speed was highlighted. It was a most moving experience for
his students, colleagues and admirers in the audience.

   The invited papers presented at the Ninth Biennial
Polymer Symposium may be conveniently divided into three
parts:

   Part I   Synthesis

   Part II   Characterization

   Part III   Properties

Part I covers the mechanism of ring opening polymerization
by free radicals (Bailey), synthesis of polyphosphazenes
(Allcock), mesomorphic polymerization (Blumstein) and
plasma polymerization (Yasuda). Part II reviews the techniques of Fourier transform infrared spectroscopy (Koenig),
carbon-13 nuclear magnetic resonance spectroscopy (Lyerla)
and x-ray photoelectron spectroscopy (O'Malley and Thomas),
all of which are powerful new tools for polymer characterization. Part III discusses the clustering phenomena in
ionic polymers (Eisenberg), the microdomain formation in
block copolymers (Kawai), the glass transition phenomenon
(Aklonis), highly oriented polymers (Porter) and the gel
entrapment systems (O'Driscoll).

   All of the topics discussed here are in the forefront
of polymer science and presented by the well known experts
in the field. There were also active discussions and comments by the participants. A poster session was held in
which a number of contributed papers were accepted. These
papers will be published elsewhere by the individual authors.

   It has been an honor and privilege for me to have been
given the responsibility of Program Chairman of this Symposium. I take this opportunity to thank the ACS Division
of Polymer Chemistry for their trust, Professor H. B.
Hopfenberg for his role as the Arrangements Chairman and
Professor S. Israel for organizing the poster session.

                                   Mitchel Shen
                                   Berkeley, California

CONTENTS

CEREMONY FOR THE PRESENTATION OF THE DIVISION OF

POLYMER CHEMISTRY AWARD TO CARL S. MARVEL

On the occasion of the presentation of the Division of
Polymer Chemistry Award to Professor C. S. Marvel on November
21, 1978, a number of formal and informal events, which accompanied the award address published in this volume, occurred
which added materially to the presentation and we hope tended
to enhance the tradition of the award. These events included
the presentation ceremonies, the introduction of the awardee,
and a panel discussion involving questions and reminiscing
about the awardee's career.

The Chairman of the Division of Polymer Chemistry, Bill
M. Culbertson in making the formal presentation of the award
expressed gratification at the excellent attendance to honor
someone that everyone admires, respects and appreciates. He
was pleased that Speed's two children, Mary Marvel from
Grosse Point, Michigan and Dr. John T. Marvel from Monsanto
in St. Louis, could be present to help honor their father.
Bill expressed appreciation to the Dow Chemical Company for
the financial support of the award and introduced Harvey D.
Ledbetter who was Dow's representative to the Biennial
Symposium.

After Paul W. Morgan gave a brief history of the award,
Bill Culbertson during the formal presentation of the award
made the following comments:

"It is not often that one has the very pleasurable job
of giving their grandfather an award. This certainly is my
opportunity since John Stille was my teacher.

Speed, it gives us great pleasure to present to you this award, which says on this plaque ...

The Division of Polymer Chemistry, Inc.
Award for
Outstanding Contributions to Polymer Chemistry
to
Carl S. Marvel
The 21st of November, 1978 - Key Biscayne, Florida

In giving you this plaque and the gift in this envelope, we recognize we can never fully express our deep appreciation for how you have given yourself to your students, your profession, this Division, the ACS, and the inspiration you are to us all."

Professor John K. Stille introduced the awardee as follows:

It is an honor for me to introduce my professor. I am going to do so, not by reciting all his accomplishments-- although I will mention a few of those--but I will tell you some stories about him that I learned yesterday afternoon. We sat down with a bottle of Old Forester, the catalyst which produced some interesting and amusing stories about a long, successful career marked with many outstanding contributions.

Professor Marvel was born near Waynesville, Illinois at about the same time as the automobile, the main difference being, however, that he is still running. He did his undergraduate work at Illinois Wesleyan, receiving the B. S. Degree in 1915. It was from this institution that he published his first paper, and obtained a very important storehouse of information through synthesizing 50 or 60 organic compounds. Thus, when he went to Illinois he was able to identify a very large number of organic compounds by smell. As a first year graduate student at Illinois, he took qualitative organic from Professor Oliver Kamm, and proceeded through the course by first identifying the unknown by smell and then making a derivative. This procedure began to bother his teacher so one day Professor Kamm gave Speed a mixture, saying "You can't tell me what this is." Marvel replied, "This is not going to be very hard; it has a low molecular weight alcohol, a fatty acid and an aromatic amine." He was correct, except that he missed the water. This might be considered the earliest use of gas chromatography in the laboratory.

Dr. Marvel received his Ph. D. degree under W. A. Noyes in 1920, but in his second and third years of graduate study, he was making organic chemicals in a preparative laboratory run by Roger Adams. This proved to be excellent synthetic experience, and the synthesized organic chemicals were sold to anyone who wanted them. During this World War I period, when Germany introduced a new chemical agent, Marvel was given the job of synthesizing the agent, such as chloropicrin or diiodoacetic ester, in 25-50 lb. batches so that the agent could be tested in gas masks. The hoods in the prep lab had no outlet, however, because when they were being installed, it was discovered that a structural beam was in the way of the ducts. The closest Professor Marvel says he came to "being done in" occurred while he was making chloropicrin from sodium hypochlorite and picric acid. The procedure was to stir the mixture in a jar, and as long as the mixture was stirring properly, it wouldn't heat up. The stirrer broke, however. Marvel put his head in the hood, got a "snoot full", and this nearly put him out.

It took him a little longer than usual to get his Ph. D. degree, because Adams got pneumonia, and Marvel had to work with Adams' students for two years. In 1920, he accepted a position on the Illinois staff at a salary of $1600 per year; this was raised to $2000 before the appointment became effective, since Kamm left the staff that year. At one time Marvel had a group of 37 graduate students partly because Marvel and Adams split up the graduate students when Carothers and Johnson left Illinois.

The secret of the success of the Illinois chemistry department was good students. Hanford, Joyce, Schreiber, Cairns, Saltzberg, and Brubaker were a few of the students that were at Illinois during Marvel's early years there. There was an exchange system with other faculty at other universities; Illinois received some of its top graduate students from Smith at Minnesota, Whitmore at Penn State, Kharasch at Chicago and Adkins at Wisconsin. In addition first rate graduate students came in from such universities as Nebraska, Miami of Ohio, and Alberta. These students always got the top grades, and consequently won the majority of the scholarships from Illinois, somewhat to the disconcern of the faculty outside of chemistry.

During this period, Carothers, Johnson and Marvel would hold seminars at Marvel's house, inviting some of the better graduate students. This group would pick some new or controversial topic, such as tautomerism or valence theory, divide the group into opposing sides and discuss the subject in detail.

During World War II, in 1940, Marvel was stuck on Adams' National Defense Research Committee which was responsible for synthetic organic work. After two years of this administrative work and some disagreement about how things ought to be done, Speed resigned. He was approached then by Dr. Bradley Dewey who wanted him to work with the Rubber Reserve Program. Marvel had had enough of Government agencies at that point, so he told Dr. Dewey he would go to work if he could have a budget of $100,000 a year. The next day, to Marvel's surprise, he had his money. This program lasted until 1955.

When Marvel "retired" (for the first time) by going to Tucson, he did research with a group of postdoctoral research associates, supported by grants or contracts from the Air Force (3 or 4 post-docs), DuPont (5 post-docs), NSF (1 post-doc) and the Southern Regional Labs, USDA (2 or 3 post-docs). As of November 1, 1978 he "retired" for the second time, drawing retirement pay, but maintaining a quarter-time appointment at the University of Arizona.

Professor Marvel was elected to the National Academy of Sciences in 1938, as an organic chemist, but maintains this would not have happened if he had been presented as a polymer chemist. He was President of the ACS in 1944, the only year no national meetings were held. I don't know how he worked that. His picture has been on the cover of C and E News at least four times that I know of. He has won a number of prestigious metals and awards, among them the Nichols Medal of the New York Section (1944), the Gibbs Medal, Chicago Section (1950), The American Institute of Chemists Gold Medal (1955), the Priestly Medal (1956), the International Award of the Society of Plastic Engineers (1964), the first ACS Award in Polymer Chemistry, the Whitco Award (1964), and the Perkin Medal of the Society of Chemical Industry (1965).

Tonight he will tell us about some of the chemistry he has been doing recently in a talk entitled, "New Thermally Stable and Processible Aromatic Polymers".

NEW HEAT STABLE PROCESSABLE AROMATIC POLYMERS

C. S. Marvel

Department of Chemistry, University of Arizona

Tucson, Arizona   85721

For a number of years we have been seeking heat stable polymers which were relatively cheaply made, processable either by solution or melt techniques, stable in air to temperatures of at least 300°C, curable by heating to give insoluble high melting and infusible products without gas formation, and stable to oxidation and hydrolysis, which would maintain their strength to at least 200°C. They should also have excellent adhesive properties toward metals, fibers, and glass cloth, and thus be useful as matrix resins for composites and for protective coatings for metals.

We developed a series of polyaromatic ether-keto-sulfones which had most of the properties mentioned except that of being cured on heating. We then started introducing units into the polymer chain which would hopefully add this feature to the properties of our polymers without detracting from the other desired properties.

Without going into detail, we did make polyaromatic ether-keto-sulfones with terminal or side chain nitrile groups[1], with terminal or side chain acetylene groups[2], with cyclophane units in the chain[3] and with disulfide units in the chain[4]. All of these substituted polyaromatic ether-keto-sulfones were relatively heat stable and could be cured by heating, but none was as stable to oxidation after curing as it had been before curing, and hence would not satisfy the use demands we had in mind.

We now have found two classes of substituted polyaromatic ether-ketones and polyaromatic ether-keto-sulfones

7

which seem to meet the requirements which we originally set
out to meet. One of these new polymers is an aromatic
ether-ketone with biphenylene units in the chain similar to
those which Stille has shown to be convenient for crosslinking some of his polyquinolines[5]. The other is a polyaromatic ether-keto-sulfone which has units of 2,2' -diph-
enylethynylbiphenyl in the chain, such as Arnold and
Hedberg[6] have found would give an internal cyclization reaction which converted their polyquinoxalines to high melting polymers which were insoluble and infusible. Our polyaromatic ether-ketone was prepared as follows:

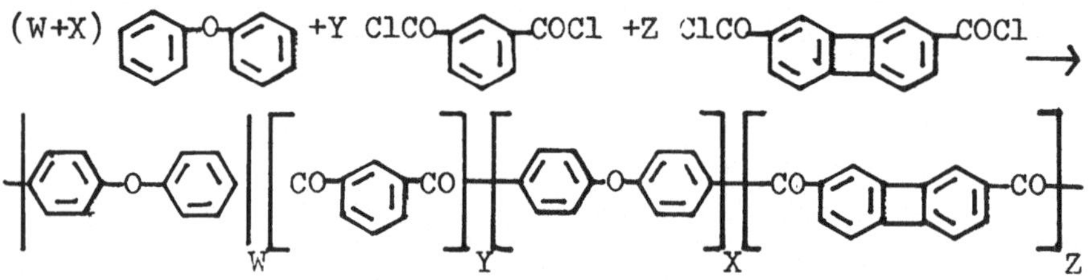

The biphenylene-dicarboxylic acid chloride was prepared and
converted to the polymer by the method of Swedo and Marvel[7].
In the synthesis were used 5 mole percent of the acid
chloride of the biphenylene derivative and 95 mole percent
of isophthaloyl derivative per each diphenyl ether. The
units of course are randomly arranged along the polymer
chain. We were able to obtain polymers of the composition
expected from the ratio of materials used in the starting
mixture. Our polymer was obtained in 95% yield with a softening point of 240-270°C. It was soluble only in sulfuric
acid and the inherent viscosity in that solvent was 0.53.
When this polymer was heated to 315°C under nitrogen for
20 hours, it became insoluble in sulfuric acid and a Vicat
softening test showed a slight softening at 120-140°C, but
no further softening below 460°. Isothermal weight loss in
an air-circulating oven at 300°C for three days resulted in
0.9% weight loss and an additional day of heating at 350°C
resulted in a weight loss of only 1%. A sample of this
polymer was cured in an I-beam mold at 300°C and 20,000 psi.
After curing, this molded product was tested on an Instron
machine and it showed a maximum strength of 11,882 psi with
5% elongation. A glass laminate was prepared from E-glass
by pressing at the softening point of the polymer and then
heating it at 300°C to cure. The laminate was strong and
not affected by any solvent tested and did not soften below

400°C. It was not brittle when cooled in a dry ice acetone
bath. The cured polymer adhered strongly to metal, and to
remove an aluminum coat it was necessary to dissolve the
aluminum in acid. Cured samples showed a Tg of about 250-
270° when tested in a differential scanning calorimeter.

A similar polymer was made by replacing some of the
diphenyl ether in the starting recipe with p,p' -diphenoxy-
diphenylsulfone. This new polymer was soluble in aprotic
solvents and had approximately the same inherent viscosity,
Tg and weight loss at 300°C as the polyketone. It did not
give as strong an I-beam on molding.

These polymers show much promise as laminating resins
to make high temperature composites, and for coating materials to be used on metal parts exposed to corrosive atmospheres. We did not attempt to make film or fibers from
these polymers, but that would be relatively easy to do if
desired.

The second promising new type of polymer was made from
2,2'-diphenylethynyl-4,4'-diphenic acid units which cure on
heating to give phenylbenzanthracene units in the polymer
chain. The new polymer also belongs to the polyaromatic
ether-keto-sulfone series. This polymer was prepared from
dimethyl 4,4'-diphenate by direct iodination in sulfuric
acid with a silver sulfate catalyst to give the 2,2'-diiodo
derivative in 80-85% yield. When hydrolized to the acid it
could be converted to the acid chloride on treatment with
oxalyl chloride in pyridine solution.

Several keto-ether sulfones were made from this acid chloride, diphenyl ether, p,p'-diphenoxydiphenylsulfone and isophthaloyl chloride in a Friedel and Crafts reaction using aluminum chloride in dichloroethane solution. The ensuing polymer with the highest viscosity (0.79 in sulfuric acid) contained one mole of the diiododiphenic acid unit, 5 mole of the p,p'-diphenoxydiphenylsulfone unit, 7 mole of the isophthaloyl unit and 3 mole of the diphenyl ether unit. This polymer was soluble in DMAC pyridine and sulfuric acid. It softened at 165°C.

The diiodo polymer was dissolved in pyridine and heated with copper phenylacetylide in a nitrogen atmosphere for 40 hours. This gave complete replacement of the iodine atoms by the phenylacetylene units.

This new polymer was soluble in DMAC, DMF, pyridine and sulfuric acid. The softening point of this polymer was 149°C. When heated to 270°C for 24 hours it cured to a flexible, tough polymer insoluble in those solvents which dissolved the uncured polymer and did not melt below 500°C.

Presumably this curing reaction involved an internal cyclization reaction which had been observed in the analogous monomer system. A thin film was made from the uncured polymer by dissolving 1 g in 5 ml of DMF, casting on a glass plate, and heating to 140°C to remove the solvent. This film was then cured at 270°C for 24 hours in a nitrogen atmosphere, when it became insoluble but was still strong and flexible. A glass laminate was made from a solution of 1 g of the polymer in DMF by pouring it on a glass cloth, draining off the excess solution and pressing four layers of the impregnated cloth at 7,000 psi and 275°C for three hours. This laminate showed no diminution in strength below 200°C. A sample of the cured polymer heated in nitrogen from room temperature to 400°C at a rate of 10°C per minute lost only 2.5% weight.

This polymer therefore shows good promise of being a
heat stable matrix resin for composites. It also shows
promise as a resin coating for metals which are to be exposed to corrosive atmospheres.

Both of these types of polymers are made from inexpen-
sive raw materials for the most part. The biphenylene di-
carboxylic acid used in the polyketone is only needed to a
very small extent of the total weight of the polymer, and
is readily obtained by literature methods from common raw
materials. Teh phenylacetylene substituted diphenic acid
used in the second polymer mentioned above is also available
by straightforward reactions from relatively cheap raw materials and is only a minor component of the final resin.
Thus, the new polymers should be capable of production at
prices that will permit their general use.

I am indebted to many excellent postdoctorate assist-
ants whose names are listed in the Bibliography for the
synthesis work in this program. I am also indebted to Mr.
George Vorie for the I-beam samples and the tests, and I am
indebted to Dr. M. Panar for some of the properties for the
cured phenylacetylene derivatives. The research was supported by Contract DAAG-29-72C-0024 of the U. S. Army
Research Office at Durham, North Carolina, and Contract
AFOSR 77-3112 with the Office of Scientific Research of the
Air Corps. The Textile Fibers Department of duPont gave us
a generous supply of dimethyl diphenate. I am very grate-
ful to all of these people for their help.

REFERENCES

1. J. Verborgt and C. S. Marvel, J. Polym. Sci. Polym. Chem. Ed., 11, 261 (1973).

2. C. Samyn and C. S. Marvel, J. Polym. Sci. Polym. Chem. Ed., 13, 1095 (1973).

3. D. M. Chang and C. S. Marvel, J. Polym. Sci. Polym. Chem. Ed., 13, 2507 (1975).

4. D. T. Wong and C. S. Marvel, J. Polym. Sci. Polym. Chem. Ed., 14, 1237 (1976).

5. J. Garapon and J. K. Stille, Macromolecules, 10, 627 (1977); A. Recca, J. Garapon and J. K. Stille, ibid, 10, 1344 (1977).

6. F. E. Arnold and F. L. Hedberg, J. Polym. Sci. Polym. Chem. Ed., 14, 2607 (1976).

7. R. J. Swedo and C. S. Marvel, J. Polym. Sci. Polym. Letters, 15, 683 (1977).

8. S. C. Lin and C. S. Marvel, J. Polymer Sci. Polym. Chem. Ed., IN PRESS.

9. E. H. White and A. A. F. Sieber, Tetrahedron Letters, 28, 2713 (1967).

PANEL DISCUSSION WITH PROFESSOR MARVEL

As part of the informal festivities of the presentation, a panel consisting of William J. Bailey, Chairman, Burton C. Anderson, John K. Stille and William E. Gibbs, after several planning sessions over a bottle of bourbon, helped Speed reminisce about his career as an organic chemist, a polymer chemist, an industrial consultant, and a government consultant. The interview which was recorded by Robert Greenley proceeded as follows:

WJB - I'd like to start off the discussion by asking Speed just how he became a polymer chemist.

CSM - Well, I think that's easy to answer. I was a farm boy and I always supposed I would be a farmer. I had an uncle who was a high school teacher who said well if you're going to college (and my mother insisted that all of the children had to go to college), at least study science because the next generation of farmers ought to be scientific farmers. So, when I went to college, the first course I took was chemistry and I never got out of it because I had a good teacher and I had so much fun doing chemistry. And that's just what I kept doing. I stayed on ... you don't realize it, but in 1911 to 1915, when I was an undergraduate, there was no chemistry in America. There was no job for anybody except maybe a few routine analysis jobs in drug companies. But all the industrial and research labs were getting stronger then. So, I didn't know how I was going to make a living, but I thought I'd have fun while I went to college and then go back and farm. In my last year,

13

things began to break in Europe. Chemicals were needed and
things got to be a little different. My professor said,
"How would you like to go to graduate school?" I said,
"What's that?" If you ever went to a small college in 1915,
you didn't have an idea what graduate work was or even what
it meant. He said, "I can get you a scholarship at the
University of Illinois if you will go." I said, "Let me
talk to my folks about it." My father said, "Well, if
somebody is going to pay for you to go to school, you'd
better go. I've been spending my money to send you." That's
how I got to graduate school. The dean of the graduate
school looked at my record and said, "How'd you ever get a
fellowship? You don't know very much." Because my college
was a small school, it didn't teach much chemistry compared
to what you have to know to do graduate work, and he gave
me a full load plus a fourth for the first year I was a
student. And I must say it was an awfully rough year, but
I lived through it and kept on being a chemist.

WJB - Well, how about telling us how you became interested
in polymers.

CSM - I was just an orthodox organic chemist doing ordinary
preps and three projects I worked on in the early 20's and
early 30's produced polymers but nobody knew what they were
at that time. And I had to find out. Actually, in 1928,
I published a paper on the first linear polyethylene that
had ever been made, I guess. But at the time, we didn't
know it was linear polyethylene; nobody was interested in
polyethylene. But since then I found out that was what we
had by polymerizing ethylene with lithium butyl. We got
it because we'd run another reaction which should have produced either ethane and ethylene or butane. The gas that
was produced was all ethane, no ethylene and no butane.
How are you going to explain that? Simple-minded me says
we must have polymerized the ethylene. I didn't know
anything about polymers, but that's what I said. So, I
took lithium butyl and Nujol, passed in ethylene and sure
enough out on top came nice solid polyethylene. That
was my first polymer. A little later I tried to make some
bromoalkylamines and these were going to make drugs for me.
But they didn't behave right. They reacted with themselves
and gave things I tried to describe and found out they
had both ionic bromine and non-ionic bromine. With that
information you could figure out that the things had polymerized and you had certain end groups with bromine

which gave you the molecular weight of those polymers. That was the next polymer I made. Then the next one I got into, I guess, was working with one that I had learned about as a consultant with DuPont. Elmer Bolton talked to me about the second time I was down there. He showed me a book of Carlton Ellis's which was the only thing written about polymers in those days. In that book was a statement that ethylene and SO_2 would give a polymer. He said, "Do you think that reaction would go?", and I said, "It doesn't seem reasonable to an organic chemist and I don't see how that would go." But he says, "Now just think, with sulfur dioxide at \$5 a ton, ethylene 2¢ a pound, if that really worked, that would be some polymer." And I said, "If you're interested, I'll go home and see if ethylene will react with SO_2." That's when I really started working on trying to make a polymer and darn if it didn't work. But nobody has really found out yet what you do with the ethylene-SO_2 polymer because it's so intractable; it won't melt; it won't dissolve; and you can't do anything with it so it never got to be any use. Some of the other olefin-SO_2 polymers have found uses in industry.

WJB - You mentioned your consulting. Burt, do you want to reminisce?

BCA - Well, in addition to the many statistics on Speed's consulting career which are in the program tonight, we should realize that a predigious amount of consulting is involved in 242 round trips from either Urbana or Tucson to Wilmington and about 3.4 years in the DuPont Hotel. That is a lot of dedication to duty to travel that much.

JKS - He told me yesterday that if he knew he was going to spend 3.4 years in the DuPont Hotel he probably would've told DuPont where to go.

BCA - Also, I think I heard Speed promise earlier this evening to consult for us for less pay in the future, but I will not take that offer for back to Wilmington. Speed, tell us how your consulting career started with DuPont and something about Wallace Carothers' role in it.

CSM - It was back in the summer of 1927, Roger Adams was the consultant for DuPont, resident consultant for the summer. At the end of the summer they wanted him to promise to come back once a month for the next year to consult. And Roger

says no, that's too much traveling on the train - I won't
do it. So, he and Charlie Stein battled back and forth.
Finally, Roger said, "Why don't you try two of us?" Stein
said, "Fine, who else you got out there?" Roger said, "I
think Marvel will do it." So that's how I got the job.
He didn't know who I was - didn't care who I was. They got
Roger to agree to come every other month. And that's
literally true - it's not as a joke. About the time I got
there and got acquainted with the new people was about the
time they decided to start Purity Hall. That's the name
the boys on the central research team gave the pure research
laboratory Stein set up around Carothers. He told Carothers
to work on anything he wanted to in research. Carothers
had been at Harvard, but he was not happy up there. Before
that he'd been with us as an instructor at Illinois - didn't
mind that too much. He went to Harvard and disliked it and
they offered him a job at DuPont. He jumped at it, and
went there. Both Adams and I had recommended him because
he's probably the smartest man either one of us ever knew
in chemistry. When he went there, he decided he was going
to work on polymers. That association, as a consultant
working with him, is another one of those things that pushed
me into more and more work in polymer chemistry. Carothers
was really a very brilliant scientist. He was an eccentric
person, a little unhappy, but he was so smart he made the
rest of us feel just a little bit inferior, except he didn't
think that way. I remember after he'd made nylon and he'd
been recognized as a member of the National Academy, he
used to complain he didn't have any new ideas and never would.
I once said to him, "My God, doc, if you think that about
yourself, what do you think about your friends?" That
straightened him out for about a year. It was inevitable
he'd get unhappy sooner or later and would take care of him-
self, and that's what happened. Really, after his paper on
polymerization appeared in Chemical Reviews, any intelligent
organic chemist could work in the polymer field without any
trouble. He just layed out all of the important things.
Staudinger had the right idea about polymers, but I don't
know if you tried to read his German papers or not. I don't
read German too well, but I tried to read them and I was still
not quite sure what Staudinger meant. He was right. He did
not care if we knew that, but he didn't sell his ideas to the
ordinary chemist. But Carothers could write so plainly, so
straight forward, that after you read his papers, you never
really questioned again about how you do it and what you're
doing.

WJB - One of your most significant roles is what you did during the World War II for the government and Bill wants to follow-up with some questions.   Bill.

WEG - Well, some of the things I knew because I was attached to the rubber reserve program from a different point of view being in Akron.   Every year there was an annual meeting which was held in Pittsburgh at Mellon Institute which was really a very unique situation with probably the best minds in chemistry meeting there and the graduate students as well were very interesting and very informative.   It was pretty exhilarating to be there.   You lived through many of those times, Speed, with Peter Debye, Paul Flory and all the rest of the people.   Do you have any reminiscences about those meetings?

CSM - Well, I can tell you earlier events.   When we started out, there were not too many of us and we use to have these meetings in our basement room and after the meetings, generally the group met in my living room.   I found out, finally, why they liked to come to Urbana so well.   I could fix old fashioneds better than anybody in the crowd.   So, working there together, we really became close to each other; I think that helped settle that whole group into a group that worked well together.   Didn't have any jealousy - didn't try to hide anything from each other.   We had to have rubber in a year and we got it.   It was an exciting time.   Then later on we had more complete meetings - bigger groups.   Then when the war was over, they sent some of us to Germany to learn what they had done.   And we again had the short meetings in Urbana to debate; you do this, you do that and get back here next months and tell us how far you've gotten so we know.   So, the cold rubber program developed very fast.   I think I was mainly the catalyst for furnishing the old fashioneds that kept them happy.   I don't think I was a great rubber chemist, but I was a pretty good mixer of drinks.

WEG - You remember that there was a meeting that I attended one time in Dayton and a good friend of ours, Art Bueche, happened to be a session chairman.   You'd given a paper on thermally stable elastomers and Art, being a student of Debye, was inclined to the theoretical quite strongly.   He was trying to probe your mind to learn what were the fundamental things you were talking about or thinking about when you sat down in the laboratory to synthesize an elastomer. Do you remember that?

CSM - I don't quite remember, but I've heard about it since. You can tell them.

WEG - Well, Art, being very eloquent as he is, stood before the group and said, "Professor Marvel, what do you have in your mind when you sit down to synthesize a new structure that should be elastomeric and have a Tg of -80° and so on? What factor do you use in constructing a model that you can set up to prepare in the laboratory?" Speed's response was, "Well, what I do is I take the best idea that I have and I run into the laboratory and make it. When it's done, I take it in my hands and I do like this (stretching motion). If it survives, it's an elastomer.

CSM - I've never been a theoretical chemist. I belong to the old family who do the compounding - chew the rubber to see whether it's right. That's what kind of chemist I am rather than a theoretical chemist. I believe in making them and having them work."

WJB - I'm going to ask you to reminisce on how you got into the government work, Speed.

CSM - Well, in 1940, I was in Canada having a wonderful time fishing when I got a telegram from Urbana to come home - you have to go to Washington on the National Defense Research Committee and help us do some work for the government. So, I did. I gave up the fishing. I haven't been fishing much since. Roger was head of the work on chemical warfare. He was in some awfully interesting chemistry under the National Defense Committee. I was given the task of setting up all the programs in organic chemistry that were being done by the National Defense Committee at that time. Well, this was the time when chemical warfare agents were studied, chemical toxicity, together with everything else that went with it. So my program developed from about the first of September in 1940, when I'd go to New York every Sunday afternoon - sit all day in a meeting at Rockerfeller Center on the toxicity program - go home that night, try to catch up on a little work the next day and leave Wednesday noon for Washington to go down for a meeting with the National Defense Research Committee on Thursday. Go home again Friday night and then back again. So, I had maybe about four nights a week in

the pullman car during that time. Being only a young NDRC
man, I didn't have enough status to get on an airplane seat
because some Colonel's dog could kick me off. So, I rode the
train. Well, you can get a little tired of that. When I
got home one night, it was a Thursday night; before I got to
bed the telephone rang - says you got to be back here Satur-
day morning to start a program for a new toxic agent from
Proctor and Gamble; want you to set it up; you got to be
here. I said I'll come and I did. We worked all day Sunday.
We got the program set up so I could go to my meeting in New
York the next day and then home. When I got home on Tuesday,
I got a call from Washington: "Unfortunately, we have'nt
got the money to set up that program." Well, that was the
end of my NDRC stuff. I was so damn mad I said I wouldn't
work for them anymore and I didn't. It didn't bother them
very much. I was in Washington swearing about that when
the Pearl Harbor episode set things off - we had no rubber.
That's when they started that program and Brad Dewey was
there. He called me and says, "You've got to come work on
the rubber program - you're the only University man who works
on polymers." I said to hell with the government. I've
had enough of their programs and I'm not going to do it.
But he kept arguing and I finally said,"Well, if they'll give
me enough money to get a good program going, I'll go to
work and I think I'll need about $100,000 a year for that."
The next morning he called and said you got it and wanted me
to go to work on Wednesday. So, I had that and then we worked
on rubber for the duration. At the end of the war, five of
us were sent to work in Germany. That was another experience
I might tell about because I'd never worked with the Army
directly. The best way to get anything done in the Army if
you have to travel is to find out who the transporation
officer is. Then you find out who the transportation officer's secretary is - then you find out what ration books you
need to get the right liquor supply in Germany - that's where
we were. When you get all that done, then you can get orders
cut for transportation and get around. Well, we could'nt
at first. In three weeks we learned only that much, and the
last week we got some work done. So about the time we got
the work done was the time when everything was over. VIP's
were no longer VIP's and any private of the last rank had
better chance of getting home than we did. I had a fellow
with me, Ed Weidlein, senior Ed's son. He was a Navy officer
who'd been put in the Navy on a special quota for his rubber
job. But he had to be home by a certain date or he could not
get out of the Navy and would have to stay the whole time as

an enlisted man.   Poor old Ed was fit to be tied and I was
a little mad about it too.   So, we got back from an all-night
trip through Germany and came back across the English Channel
in a short boat.   It was so damn rough , the ocean, that the
Queen Mary wouldn't go out.   That was a pretty rough trip.
We got home and once I got through our office there, I didn't
see any way for me to get home.   Since they sent us back, we
were just ordinary people and I would say a simulated colonel
(that's what they called us) fell just below the last private
in the rear rank.   We were feather merchants as far as the
Army was concerned and they weren't going to take care of us.
I got hold of somebody in charge of transportation, a fellow
from Chicago, Halsey Stuart, who was the transportation officer for this place.   So I thought I'd try my trick on him -
I'd tried it on his secretaries.   So, I said to him, "Officer
Stuart, do you know my friend Roger Valentine?"   Yes, he knew
him.   "Well," I said, "come on, let's go have a drink."   We
went up to the Officers Club and I gave him enough Irish
Whiskey that he was drunk enough that he promised to send
us home.   And I was still sober enough to see that he kept
that promise.

        During the war I also had a turn in the Malaria Program
in the Committee on Medical Research.   I tried to get out
of that one, too.   There I was not a chemist but only a man
to get other people to work, a whipcracker.

WJB - Well, you were telling us earlier a few other drinking
stories about how you got on the train and you had to plan
ahead.

CSM - Oh yeah - in the old days when you had to go to New
York on Sunday night, prohibition was everywhere except the
state of Illinois.   So, when we got on the train in downtown
Chicago, we rushed to the bar to order all we thought we
could drink that night.   We'd sit out front until we got it
used up.   It was a trick you learned; you knew these things
were necessities sometimes.

JKS - I was going to ask you about the legendary Farwell's
which I guess you could describe as a greasy spoon, but it's
a place where we graduate students always used to go to get
coffee.   As graduate students, we would sit and watch you
people.   Rumor had it that you were running all the business
of the department from the coffee table.   We also thought

you were making up comprehensive questions or preliminary examination questions, but you, Roger Adams, Bob Fuson, Dave Curtin, and Harold Synder used to hold some conclaves. What really went on?

CSM - First thing that always happened was that you got out your coin and flipped it. Heads was out. If you got heads, you were all right - the rest of them bought the drinks.

JKS - You never bought coffee, did you?

CSM - Not very often! For a long time I had a two-headed coin. Till they caught on to it. I remember too, that all the time I was in charge of the organic division at Illinois - don't know whether I was ever in charge of it officially, but I was actually - I didn't have any such title - we never had a faculty meeting - except over at Farwell's. We'd sit down, have a drink and talk about what ought to be done. My policy was don't talk with a bunch of people - go out and see each one of them - find out what they want and then try to decide what ought to be done and go back and tell the rest of them that's what we're going to do. It worked pretty well. Nobody seemed to get upset about it. I think faculty meetings are the worst thing in the world. I never had any faith in them and I don't think much about committee meetings. Better to try to do for yourself without too much jawboning about it. It was just business as usual.

WJB - My favorite story was you passing the exam that Alexander gave.

CSM - Well, Elliott Alexander was one of the very bright young men in his field. He thought this mechanism chemistry was really important and if you knew how to push or pull the electrons, you could do anything. I always kidded him and told him you had to know the answer first to know whether you wanted to push or whether you wanted to pull. I still believe that. Elliott went through his course and at the end he pulled out one of these fancy examinations in which he asked the people to partake. What was going to happen in a bunch of cases that were set up so that the students were supposed to push or pull the electrons to get the right answers. So he brought it in and showed it to me. As I wrote out the answers, I said, "That's a pretty easy exam. They should get the right answers." He said, "How in the

hell can you do that when you don't know what an electron
is?" But the truth of the matter is that I still believe
if you know the chemistry, you can still tell where they're
going without having to know how the electrons behave. You
may not realize it, but when I started studying chemistry,
my professor was a student of Kahlenberg. He was a great
inorganic chemist at the University of Wisconsin and he was
the last of the die-hards who wouldn't believe that such a
thing as ions existed in solutions. I grew up with a little
bit of feeling that many of these ... theories would probably be exaggerated and you shouldn't take them seriously.

BCA - Speed, I just want to make a comment - really not a
question. I want to pass on the good wishes of all your
many, many friends at DuPont including the 40 graduate
students we have whom you trained, and your many other
postdoctoral students. I know they all want me to pass on
their congratulations in receiving this award and we look
forward to seeing you again next month.

CSM - I'll be there - I hope.

WJB - Yes, a lifetime contract with DuPont.

WEG - I'm almost afraid to ask this, but I think the group
assembled here should hear the honest story how the nickname "Speed" came up. I know most of us have heard it, but
I think the group assembled would be interested in how
that happened.

CSM - Well, that happened when I went to the University of
Illinois. It was 1915, in the fall, that I moved into Gamma
Alpha house. I don't know whether you people know what that
is or not. It's a graduate scientific fraternity which in
those days was really a good bunch of boys. We had all the
graduate students and all the scientists living there. It
was not just chemistry. There was one fellow, who was a
zoologist, who thought everybody ought to have a nickname.
He was worried about me for a long time and the fact that
I could get up later and make breakfast on time more than
anybody in the the house. That's where my name came from -
I never hurried any other time.

JKS - I wanted to say that I know earlier this evening you
said you borrowed an idea from me. I was just trying to pay
you back is all. And I never will completely, because the

ideas and the inspiration on the kind of chemistry that you've done and the contributions that you've given to my own research are much more than I can ever repay.

CSM - Well, thank you, John.

WEG - Tonight on the panel I also represent the group of people who were not fortunate enough to be graduate students of Speed's. You adopted me back a few years ago so I feel a certain kinship. But, certainly in the difficult areas like thermally stable polymers, I've had the opportunity to see you in action and work along with you in many cases. It's been a source of inspiration to all of us to see how good chemistry can be done and share a bottle of Wild Turkey now and then and have a good time doing it.

CSM - Well, I think that's important, too. Chemistry should be fun and if it isn't fun, by gosh you're in the wrong profession. That's all there is to it. As a matter of fact, I've talked some students out of it - I've even talked two of them into being preachers.

WEG - They're not having nearly as much fun as you are.

CSM - That's right, you know.

WJB - There's one aspect in your life that hasn't come up at all tonight, and this is bird watching.

CSM - I've been doing that here.

WJB - We can't let this go by without making some comments on the one thing you carry with you all the time.

CSM - I always have my binoculars wherever I go. I've always been interested in the outdoors. I was a fisherman for a long time and interested in birds at that time too. When this NDRC came along, I had to give up fishing but I could do bird watching wherever I was, including - hell, what was the name of that place that we worked in Washington all the time? There was a big garden out there. I can't remember the name of it. Oh, yes, Dumbarton Oaks. They had a nice bunch of trees and parks around there. I could sit in some of those dull meetings and watch out the windows. So, I had more fun being a bird watcher. I found out I could do that wherever I was -- in town, or out of town. As I

kept working, Bob Frank got me interested in this. I got
to keeping a list trying to see how many I could get.
Finally, I got 625 North American birds which puts me in
the 600 class. We brag a lot. Wherever I go, I do it.
This reminds me of the trip we were going to Moscow. They
asked me to come over and give a talk at a IUPAC meeting.
I said yes, I'll go. If I came, I wanted someone to take
me bird watching over there because that's what I'm really
interested in, not the IUPAC meeting, but I'd come give
a talk. Oh, yes, they'd do that. They had someone over
there just as big a liar as I was. I went out bird watching
by myself and everytime I picked up the binoculars, I
wondered whether the guy behind me was going to pick me off.
But they never did.

WJB - I know one of your favorite publications was the one
on the mating habits of the Cape May Warbler.

CSM - Yes.

WJB - When was that published?

CSM - Oh, I can't remember that. I don't remember. Another
one I got was the occurrence of the blue grosbeak in Ontario
where it was over 1,000 miles north of where it was supposed
to be. I found out since though, that the birds don't read
the books. Well, since I got here, I sit in my front yard
down there and I've seen four that I haven't found anywhere
else this year. So you can get them anywhere you go and
there's a lot of them.

WJB - Art Cope used to tell some of the stories about when
you were up at Cliff Lake.

CSM - Well, maybe some of those are better not to be told.
We really had a great time. We, there was a bunch of chemists
used to come up there to visit. There was Art Cope, Walt
Kirner, Ernie Volviler and Cliff Hamilton. We really had
a wonderful time and good fishing too.

WJB - You were catching muskies out there - forty pounders?

CSM - Well, that's a little bit high. They weren't over 40
inches. We got good muskies - some years we got them, some
years we didn't. I always said if you were going to Canada
for fish that you fish for walleyes for food, northern for

fun, and muskies to brag about when you got home.

JKS - About the only thing that we haven't covered are some
of the poker games. What was the most interesting poker
game you've ever been in?

CSM - I think the most interesting one was one that we played
one night at Dick Schreiber's house in Wilmington, Delaware.
I don't remember who all was there, but I know Dick was, Bob
Joyce, of DuPont, who was a very outspoken gentleman and who
was a very good poker player who hadn't won a damn thing. I
don't think he even won a pot. His luck was completely bad.
When the game broke up, everyone threw their odd chips into
the ring - maybe about 50¢ in there. We dealt a cold hand
and Bob Joyce got a straight flush dealt cold. Couldn't
raise on it. Couldn't do anything. He was fit to be tied.
That's one game I'll never forget. Another one I remember,
we played down in Waynesboro. I don't remember all of that
crowd. I went down there one time just before Christmas
and the boys said will you play with us tonight. I said,
"Yeah, I'll play a game or two." One of them said you gotta
be careful, it's too close to Christmas and they're all broke
and want money. All right, I said, I'll be careful. And a
little later he came back and said, "God, there's some crazy
rules down here. You get a good straight or good flush and
the next hand is a wild one; you have a wild hand and any
kind of game you want to play, and there's no limit. Or
maybe a 50¢ limit." Well, I played along that evening and
I had one hand, and the first five cards I got, including
wild cards, was five kings. Naturally, when it came to my
hand, I raised and when we finally cashed in the pot, I'd
won $7.50 that night and it's the only hand I won.

WJB - Are there any other questions or do you want to make
a statement, Speed?

CSM - I'd like to say again that I don't think anybody has
had more fun being a chemist than I have. I don't think
there's any better group of people in the world than the
chemists. They're the tops as far as I'm concerned. They're
fun to be with and most of them are not too serious - just
serious enough to do their job, and as far as I'm concerned
they can't be beaten. I hope I can still do chemistry for
another year or two at least. You can't last forever, you
know that, but I hope I will for a while yet.

PART I
SYNTHESIS

FREE-RADICAL RING-OPENING POLYMERIZATION[1,2]

William J. Bailey, Paul Y. Chen, Wen-Bin Chiao, Takeski Endo, LuAnn Sidney, Noboru Yamamoto, Noboru Yamazaki, and Kazuya Yonezawa

Department of Chemistry University of Maryland, College Park, Md. 20742

Free-radical ring-closure polymerization has been a common phenomenon in polymer chemistry for some time. Inter-intramolecular polymerization dates from the publication of Butler and Angelo[3] in 1957, in which diallyldimethylammonium bromide was polymerized by a free radical mechanism to produce a soluble polymer with the formation of five-membered rings. Apparently the reaction is kinetically controlled since the formation of the five-membered

29

ring does not produce the most stable radical intermediate. This reaction has been extended to a large variety of diolefins and is now a commercially important reaction.

Although the ionic ring-opening polymerization of heterocyclic compounds, such as ethylene oxide, tetrahydrofuran, ethyleneimine, β-propiolactone, and caprolactam, as well as the Ziegler-Natta ring-opening polymerization of cyclic olefins, such as cyclopentene and norbornene, are well known, free radical ring-opening polymerizations are rather rare. The few examples that are reported in the literature involve cyclopropane derivatives or highly strained bicyclic olefins. For example, Takahashi[4] studied the free radical polymerization of vinylcyclopropane and reported that the cyclopropane ring opened to give a polymer containing about 80% 1,5-units and about 20% of undetermined structural units. Apparently the radical adds to the vinyl group to give the intermediate cyclopropylmethyl radical which opens at rate faster than the addition to the double bond of another monomer. Somewhat similar results were obtained with the chloro derivatives.

$$CH_2 = CH - CH - CH_2 \xrightarrow{\text{AIBN}} \left[CH_2 - CH = CH - CH_2 - CH_2 \right]_x$$

(80% 1,5-units; 20% unknown)

$$R - CH_2 - CH - CH - CH_2 \longrightarrow R - CH_2 - CH = CH - CH_2 \cdot$$

↓ R°   ↑ repeat

More recently Cho and Ahn[5] studied the related malonic ester derivative, which underwent ring-opening free-radical polymerization to produce a high molecular weight polymer containing only the 1,5-units. In this case the two carboethoxy groups stabilize the intermediate ring-opened radical so that little or no addition takes place involving the internal double bond of the polymer.

$$CH_2{=}CH{-}CH{-}C{-}(CO_2Et)_2 \xrightarrow[\substack{55° \\ 72\%}]{AIBN} {+}CH_2{-}CH{=}CH{-}CH_2{-}\underset{\underset{CO_2Et}{|}}{\overset{\overset{CO_2Et}{|}}{C}}{+}$$

$$[\eta] = 2.42$$

↓ R°   ↑ repeat

$$R{-}CH_2{-}\overset{o}{CH}{-}CH{-\!\!-\!\!-} C{-}(CO_2Et)_2 \longrightarrow R{-}CH_2{-}CH{=}CH \quad {}^{°}C(CO_2Et)_2$$

Errede[6] found the dimer of o-xylylene would undergo free-radical ring-opening polymerization to give the corresponding poly-o-xylylene. In this case the driving force for the ring-opening step is the formation of the aromatic ring.

↓ R°   ↑ repeat

The course of some of these ring-opening and ring-closing polymerizations can be explained by the recent data of Maillard, Forrest and Ingold[7] that are listed in Table I.

Table I

	k_{250}, s^{-1}	E, kcal/mol	$\log A/s^{-1}$

$$\begin{array}{c} CH_2 \\ | \\ CH_2 \end{array}\!\!\searrow\! CH\!-\!CH_2{}^{\circ} \longrightarrow \begin{array}{c} CH_2{}^{\circ} \\ | \\ CH_2 \end{array}\!\!-\!CH\!=\!CH_2 \qquad 1.3 \times 10^{8} \qquad 5.94 \qquad 12.48$$

$$\begin{array}{c} CH_2 \\ \| \\ CH \\ \diagup \\ CH_2 \quad CH_2{}^{\circ} \\ | \qquad | \\ CH_2 \!-\! CH_2 \end{array} \longrightarrow \begin{array}{c} \overset{\circ}{C}H_2 \\ | \\ CH \\ \diagup \quad \diagdown \\ CH_2 \quad CH_2 \\ | \qquad | \\ CH_2 \!-\! CH_2 \end{array} \qquad 1.0 \times 10^{5} \qquad 7.8 \qquad 10.7$$

They studied the transformations in the cyclopropylmethyl and the cyclopentylmethyl series by electron spin resonance. In the case of the three-membered radical the reaction in- volves ring-opening since the energy is favorable and the rate of the reaction is very high. In the case of the five- membered ring system the reaction proceeds in the direction of ring-closure since the energetics of that reaction is favorable and the rate of the ring closure is also moder- ately high.

In a research program designed to find monomers that would undergo free-radical ring-opening polymerization, it became clear that the polymerization would be successful only if the ring-opening process was favored energetically and that the activation energy for the ring opening was low so that the rate for the ring-opening process was much fas- ter than the propagation step of the polymerization. It was reasoned that if heteratoms were introduced in the ring, the energy of the formation of a heteratom double bond might be used to counteract the energy of the closure to five- and six-membered rings. Thus it was possible to write three different general classes of monomers that might undergo free-radical ring-opening polymerization.

Case I

$$\sim\!\!\sim\!\! D^{\circ} + A = \underbrace{B\!-\!C\!-\!D} \longrightarrow \sim\!\!D\!-\!A\!-\!\underbrace{\overset{\circ}{B}\!-\!C\!-\!D} \longrightarrow \sim\!\!D\!-\!\underbrace{B = C \quad D^{\circ}}$$

Case II

$$\sim D° + A = B-C-D \longrightarrow \sim D-A-\overset{°}{B}-C-D \longrightarrow \sim D-A-B = C \quad D°$$

Case III

$$\sim D° + A-B-C-D \longrightarrow \sim D-A-\overset{°}{B}-C-D \longrightarrow \sim D-A-B = C \quad D°$$

In the case of the polymerization of the vinylcyclopropane derivatives, the process falls into Case II in which the ring-strain of the three-membered ring promotes the cleavage of the C-D bond. In the case of the polymerization of o-xylylene dimer, the process also falls into Case II but in this case the formation of the very stable aromatic unsaturation between B and C produces the driving force for the ring-opening step. It appeared quite likely that there would be many other unsaturated monomers in which the double bond formed between groups B and C would be sufficiently stable to promote the ring-opening reaction. For example, if C was an oxygen atom and B was a carbon atom, in either Case I or Case III, it was reasoned that the formation of the stable carbon-oxygen double bond might furnish enough driving force to promote the ring-opening even in a five-membered ring. In fact it has been estimated that a carbon-oxygen double bond is 50-60 kcal. more stable than the comparable carbon-carbon double bond.

A search of the literature indeed revealed many ring systems containing an oxygen atom that would undergo a ring-opening reaction in the presence of free radical catalysts.

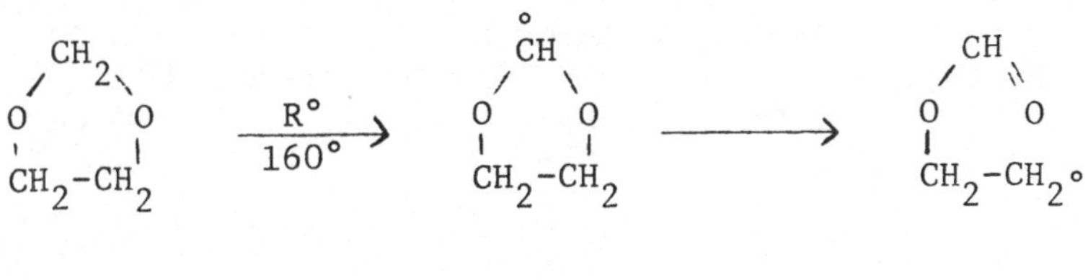

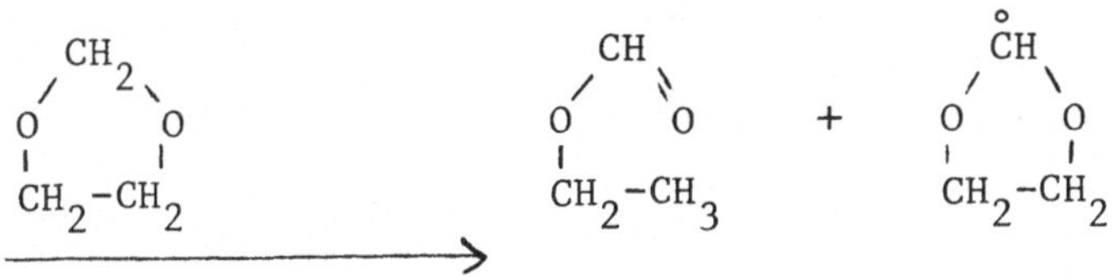

One such case was the cyclic formal, ethylene formal.
Maillard, Cazaux, and Lalande[8] found that when ethylene
formal was heated at 160°, it rearranged to ethyl formate.
The reaction could be rationalized as indicated where the
driving force for the ring-opening reaction in the chain
reaction was the formation of the stable carbon-oxygen
double bond in the final ester. With the knowledge that
such a ring system would undergo cleavage, it seemed to be
a fairly straight forward process to synthesize a monomer
that would undergo ring-opening polymerization by introducing a double bond at the carbon atom flanked by the two
oxygens.

McElvain and co-workers[9] had prepared a large number
of ketene acetals but had never prepared ethylene ketene
acetal. We, therefore, decided to prepare this monomer by
the general method of McElvain by the following set of reactions from the commercially available bromoacetaldehyde
diethyl acetal. Treatment of this monomer with benzoyl
peroxide gave a high molecular weight polyester by a free-
radical ring-opening polymerization which can be rationalized by the accompanying scheme. The structure of the polyester was established by analysis as well as infrared and
nmr spectroscopy. This polyester is difficult to synthesize with high molecular weight from the γ-hydroxybutyric
acid because of the stability of the competing lactone.

One of the intriguing characteristics of the ethylene
ketene acetal is its ability to copolymerize with a wide
variety of common monomers, including styrene and methyl

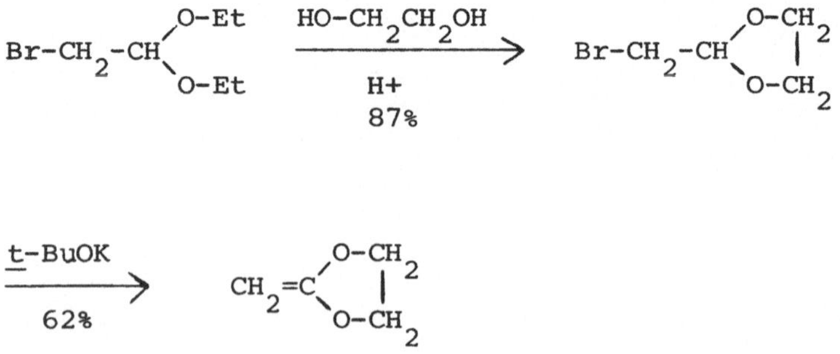

methacrylate.   One should note that this process introduces
an ester group into the backbone of an addition polymer.
Although the copolymerization of oxygen would introduce a
peroxide linkage into the backbone, this is the first time
that a relatively stable but yet hydrolizable functional
group has been introduced into the backbone of an addition
polymer. The use of this type of reaction to prepare more
thermally stable polymers, biodegradable polymers, and functional polymers will be discussed later in this paper.

$$\text{(structure: tetrahydrofuran-CH}_2\text{OH)} \xrightarrow[\substack{49\%}]{\substack{\text{Na then NH}\\ \text{CH}_3\text{O-C-C}\equiv\text{N}}} \text{(structure: tetrahydrofuran-CH}_2\text{-O-C(=NH)-OCH}_3)$$

$$\xrightarrow[\substack{\underline{\text{n-PrNH}}_2\\29\%}]{495°} \text{(structure: 2-methylenetetrahydrofuran, =CH}_2) \xrightarrow[130°]{(\text{t-BuO})_2} \left[\text{CH}_2\text{-C(=O)-CH}_2\text{-CH}_2\text{-CH}_2 \right]_x$$

repeat

$$\downarrow \text{R}°$$

$$\text{(structure: tetrahydrofuran-CH}_2\text{R)} \longrightarrow \text{(structure: cyclic radical intermediate)}$$

Previously we were able to prepare 2-methylenetetrahydro-
furan[10] by the above scheme.   Many previous attempts to
synthesize this monomer by either dehydrohalogenation or
pyrolysis of common esters failed either because of the high
reactivity of the methylene derivative or because of the
very high temperature required for pyrolysis.   However, the
imidocarbonates decompose at a temperature nearly 150° lower
than the corresponding acetate.   Even so n-propylamine was
added to tie up the liberated cyanic acid to prevent acid-
catalyzed rearrangement of the product.   Polymerization gave
the ring-opened polyketone whose formation can be rationalized by the addition of a radical to the vinyl ether to produce the somewhat hindered tertiary radical that in turn can
undergo ring-opening.   The driving force for the ring-opening process is undoubtedly the formation of the stable carbon-oxygen double bond.   The copolymerization of such a monomer with a variety of comonomers represents a convenient
way to introduce a carbonyl group into an addition polymer
and may be of some use in producing photodegradable polymers.

In a program to develop monomers that expand on polymerization, we had prepared a series of spiro orthoesters, spiro
orthocarbonates, trioxabicyclooctanes, and ketal lactones
that could be polymerized with ring-opening by ionic catalysts to give either no change in volume or slight expansion
in volume.[11]   Since a large proportion of industrial polymers are prepared by free-radical polymerization, it was

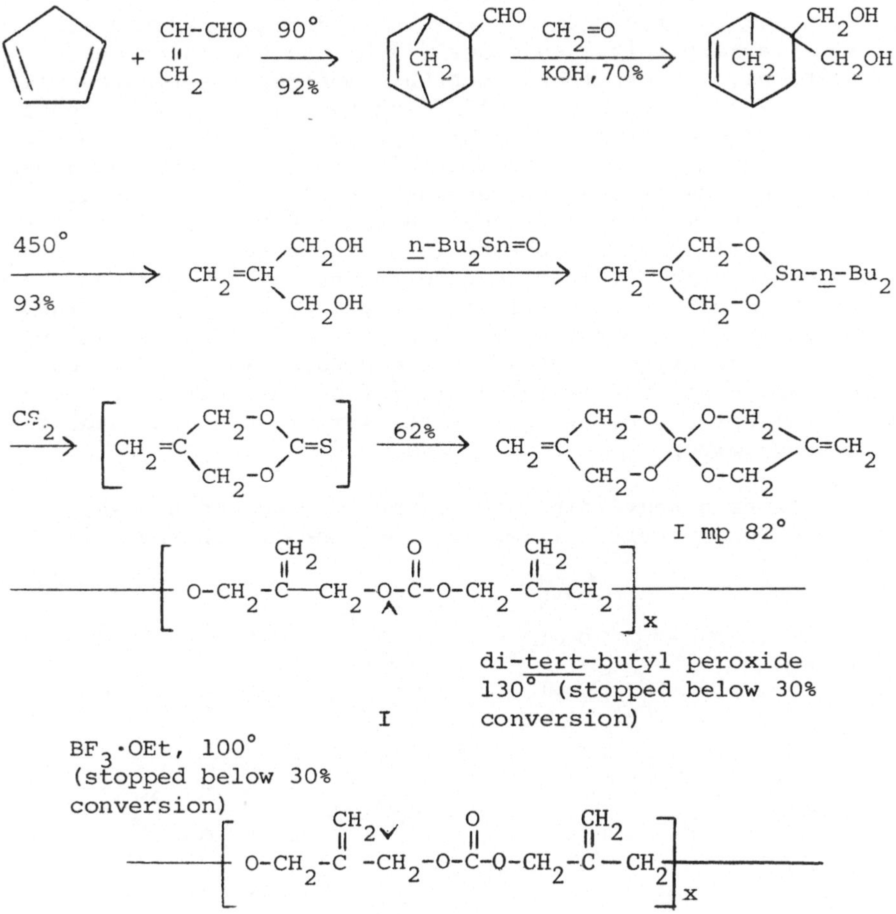

$$\text{(cyclopentadiene)} + \overset{\text{CH-CHO}}{\underset{\text{CH}_2}{\overset{\|}{\,}}} \xrightarrow[92\%]{90°} \text{(bicyclic-CHO, CH}_2) \xrightarrow[\text{KOH, 70\%}]{\text{CH}_2=O} \text{(bicyclic-CH}_2\text{OH, CH}_2\text{OH)}$$

$$\xrightarrow[93\%]{450°} \text{CH}_2=\text{CH}\overset{\text{CH}_2\text{OH}}{\underset{\text{CH}_2\text{OH}}{\,}} \xrightarrow{\underline{n}\text{-Bu}_2\text{Sn}=O} \text{CH}_2=\text{C}\underset{\text{CH}_2-\text{O}}{\overset{\text{CH}_2-\text{O}}{\diagdown\diagup}}\text{Sn-}\underline{n}\text{-Bu}_2$$

$$\xrightarrow{\text{CS}_2} \left[\text{CH}_2=\text{C}\underset{\text{CH}_2-\text{O}}{\overset{\text{CH}_2-\text{O}}{\diagdown\diagup}}\text{C=S} \right] \xrightarrow{62\%} \text{CH}_2=\text{C}\underset{\text{CH}_2-\text{O}}{\overset{\text{CH}_2-\text{O}}{\diagdown\diagup}}\text{C}\underset{\text{O-CH}_2}{\overset{\text{O-CH}_2}{\diagup\diagdown}}\text{C=CH}_2$$

I mp 82°

$$-\left[\text{O-CH}_2-\overset{\overset{\text{CH}_2}{\|}}{\text{C}}-\text{CH}_2-\text{O}-\overset{\overset{O}{\|}}{\text{C}}-\text{O-CH}_2-\overset{\overset{\text{CH}_2}{\|}}{\text{C}}-\text{CH}_2 \right]_x -$$

di-tert-butyl peroxide
130° (stopped below 30%
conversion)

I

BF$_3$·OEt, 100°
(stopped below 30%
conversion)

$$-\left[\text{O-CH}_2-\overset{\overset{\text{CH}_2}{\|}}{\text{C}}-\text{CH}_2-\text{O}-\overset{\overset{O}{\|}}{\text{C}}-\text{O-CH}_2-\overset{\overset{\text{CH}_2}{\|}}{\text{C}}-\text{CH}_2 \right]_x -$$

desirable to have available a series of monomers that would
polymerize by a free radical process with no change in volume or slight expansion. Such monomers could be added to
common monomers to produce copolymers with reduced shrinkage
or expansion in volume. It was reasoned that the introduction of unsaturation into a spiro orthocarbonate would permit double ring opening with expansion in volume. For that
reason we undertook the synthesis of 3,9-dimethylene-1,5,7,
11-tetraoxaspiro[5.5]undecane (I) by the preceding set of
reactions. It was found when this monomer was treated with
di-tert-butyl peroxide at 130° and the reaction was stopped
below 30% conversion, a soluble polymer was obtained having
a structure of a polycarbonate with pendant methylene groups.
The structure of the polymer was established by elemental
analysis as well as infrared and NMR spectroscopy. A very
similar polymer could be attained by treatment of the monomer with boron trifluoride etherate at low conversions. The
mechanism of the polymerization appeared to involve a radical double ring-opening according to the following mechanism.[13] The driving force for the double ring-opening polymerization apparently is the relief of the strain at the
central spiro atom as well as the formation of the stable
carbonyl group.

At high conversions this monomer produced a highly cross-
linked resin, very similar in appearance to the material pro-

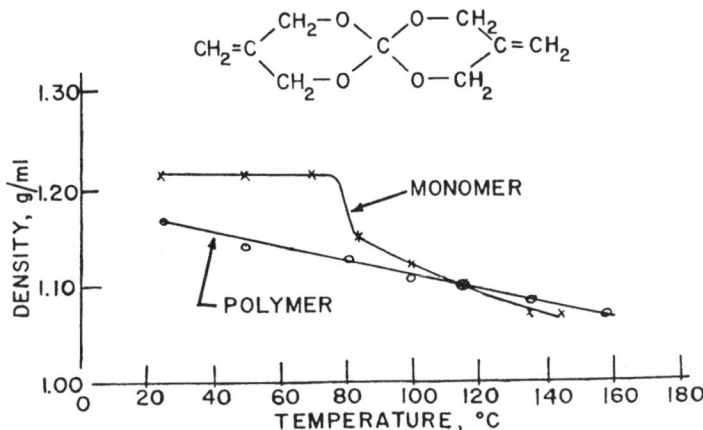

Figure 1.   Densities of the monomeric unsaturated spiro
ortho carbonate and related polyoxycarbonate vs. temperature.

duced from the polymerization of diallyl carbonate.   As
indicated in Figure 1 the volume change that occurred during homopolymerization was quite unusual.   At room temperature, a 4.3% expansion in volume occurred, while just below
its melting point at 70°, a 7% expansion in volume occurred;
at 85° a 2% expansion took place; and the expansion decreased
until at 115° no change in volume took place during polymerization; above 115° a slight shrinkage occurred.   It is
obvious from these data that the large expansion in volume
that occurs below the melting point involves not only the
increase in volume due to the double ring-opening, but also
a change in volume of 3-6% due to the process of going from
a crystalline monomer to a liquid monomer.   Since the monomer is a crystalline solid, it is difficult to find examples
of homopolymerization in which the full 7% expansion in
volume can be utilized.   However, in copolymerizations it
is possible to use a slurry of the crystalline monomer in
a liquid monomer so that as copolymerization progresses,
the crystalline monomer dissolves with some expansion and
also polymerizes with expansion.

The dimethylene spiro orthocarbonate I was shown to copolymerize with a variety of monomers, including methyl
methacrylate, hydroxyethyl methacrylate, styrene and diallyl
carbonate.   With methyl methacrylate I was shown to be less

$$CH_2=C \begin{matrix} CH_2-O \\ CH_2O \end{matrix} C \begin{matrix} O-CH_2 \\ O-CH_2 \end{matrix} C=CH_2 \ + \ CH_2=C \begin{matrix} CO_2CH_3 \\ CH_3 \end{matrix} \xrightarrow{peroxide}$$

reactive with reactivity ratios of $r_1=0.87$ and $r_2=16.4$. Since $r_1 \cdot r_2$ is greater than 1, this fact was interpreted as involving association of I in the monomer mixture rather than existing in a homogenous solution. In this copolymerization if the monomer mixture contains 10% I and 90% methyl methacrylate and the reaction is stopped at 69% conversion, the copolymer contained only 1% of the carbonate units.

As discussed earlier the free-radical ring-opening copolymerization of the dimethylene monomer I with methyl methacrylate introduces a functional group into the backbone of an addition polymer. It was obvious that the introduction of even a small amount of this unit would greatly affect the chemical and physical properties of the base polymer. For example, polymethyl methacrylate is somewhat thermally unstable with the mechanism of decomposition involving a depolymerization process. It has been shown that, when a break occurs in the chain to form a radical above 225° C, reverse polymerization will take place to convert essentially the entire polymer molecule to monomer. Since the introduction of ethylene oxide units into polyformaldehyde produces increased thermal stability, it was reasoned that the introduction of the ring-opened units into polymethyl methacrylate would also result in increased thermal stability. Thus if a break occurred in the polymer chain, depolymerization would proceed until the radical end reached a ring-opened unit. Since it is unlikely that the ring would reclose, and there is no other simple process for the removal of this end unit, the unzipping process

will stop and further decomposition of that polymer molecule
will cease. The fact that this prediction actually occurs
is shown in Table II.

Table II
Thermal Stability of Copolymers of Methyl
Methacrylate and Dimethylene Spiro Orthocarbonate

Amount of DSOC in Copolymer, %	Weight of Polymer Residue After 30 Minutes at 225° C Under N_2, %
0	78.6
1	89.8
3	94.9
8	96.3

The fact that the presence of 1% of the ring-opened unit in
the copolymer reduces the weight loss to less than one-half
that of the homopolymer is very encouraging.

Since the carbonate group is a relatively easy group
to hydrolyze, several other uses of its copolymers were sug-

gested. A copolymer of the dimethylene monomer I and styrene
was hydrolyzed to produce a hydroxy-terminated polystyrene.
The use of other monomers that polymerize by ring-opening
should permit the synthesis of a variety of addition polymers containing a wide assortment of functional endgroups.

Still another use of the presence of a functional group
within the backbone of an addition polymer is to render the
copolymer biodegradable. When the dimethylene monomer I and
hydroxyethyl methacrylate (HEMA) were copolymerized to produce a copolymer containing 14 mol-% of the ring-opened units,
hydrolysis of the copolymer in an alcoholic solution containing 1% sodium hydroxide gave within 3 hrs. at room tempera-

ture a copolymer with 1/5th of the original viscosity average molecular weight. Preliminary results indicate that
this copolymer is biodegraded by microorganisms. For use in
the human body complete biodegradability may not be necessary
since polymeric materials with molecular weights below 5,000
can be eliminated from the body. Although some hydrophilic
character is necessary for fast biodegradation, the introduction of a sufficient number of such functional groups in the
backbone may render even hydrophobic polymers, such as polystyrene, slowly biodegradable.

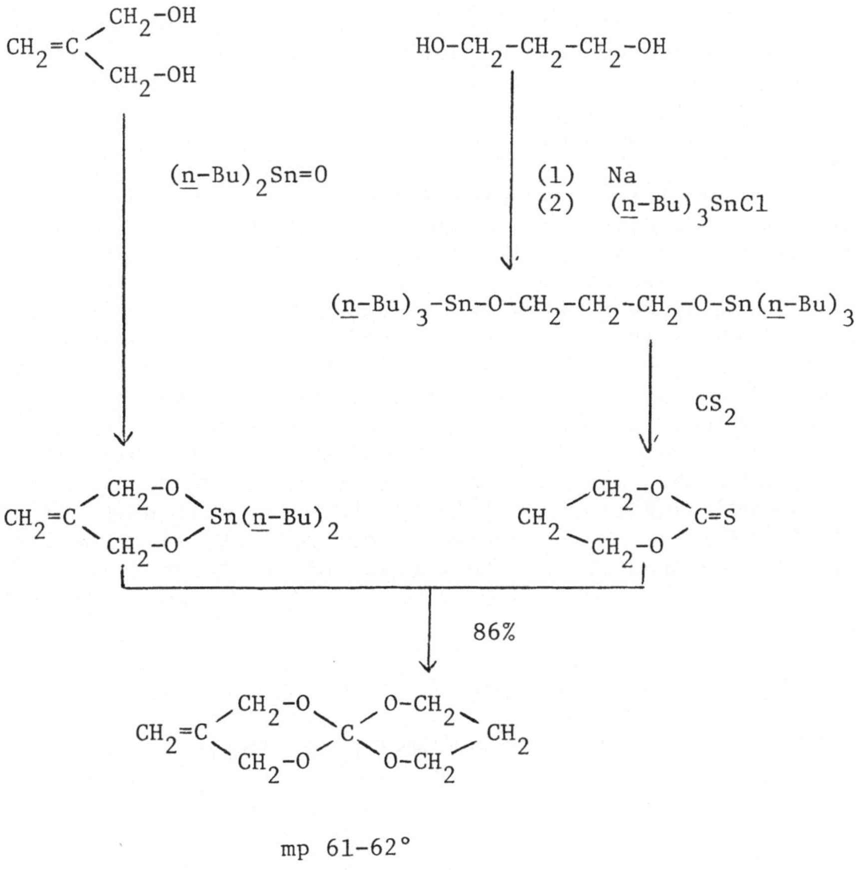

$$CH_2=C\begin{smallmatrix}CH_2-OH\\ \\CH_2-OH\end{smallmatrix}$$

$$HO-CH_2-CH_2-CH_2-OH$$

$$(\underline{n}-Bu)_2Sn=0$$

(1)   Na
(2) $(\underline{n}-Bu)_3SnCl$

$$(\underline{n}-Bu)_3-Sn-O-CH_2-CH_2-CH_2-O-Sn(\underline{n}-Bu)_3$$

$$CS_2$$

$$CH_2=C\begin{smallmatrix}CH_2-O\\ \\CH_2-O\end{smallmatrix}Sn(\underline{n}-Bu)_2$$

$$CH_2\begin{smallmatrix}CH_2-O\\ \\CH_2-O\end{smallmatrix}C=S$$

86%

$$CH_2=C\begin{smallmatrix}CH_2-O\\ \\CH_2-O\end{smallmatrix}C\begin{smallmatrix}O-CH_2\\ \\O-CH_2\end{smallmatrix}CH_2$$

mp 61-62°

II

A potential use of this monomer is in the area of dental fillings in which a slurry containing 20% of very fine
crystals of the unsaturated spiro ortho carbonate I in 60%
of the adduct of methacrylic acid to bisphenol-A diglycidyl
ether (Bis-GMA) plus 20% trimethylolpropane trimethacrylate
produces on polymerization a material with essentially no
change in volume. An investigation of a bubble test on tooth
enamel showed that this copolymer had nearly double the adhesion to the tooth structure that the base resin had without
the addition of the unsaturated spiro ortho carbonate. The
copolymer also had improved impact strength but yet essentially the same modulus, and filled composites appeared to
have somewhat improved abrasion resistance.

Since the synthesis of the spiro ortho carbonates through
the tin compounds could be modified to produce unsymmetrical
materials, we undertook the synthesis of the unsymmetrical
2-methylene-1,5,7,11-tetraoxaspiro[5.5]undecane by the accompanying set of reactions.

The resulting monomer was a crystalline solid with a
melting point of 61-62°. When the polymerization was carried
out in the presence of di-tert-butyl peroxide and the reaction was stopped at low conversion, a linear polycarbonate
containing pendant methylene groups was obtained. The structure of the polymer was established by elemental analysis as
well as infrared and NMR spectroscopy. The structure of this
material was very similar to the polymer that could be obtained by the ionic polymerization of this same monomer at
low conversions. Bulk polymerization of II with peroxide
catalyst gave a material at 25° with an expansion of 4.5%
and at 60° an expansion of 5.5%; above the melting point of
II (61-62°) the expansion decreased until at 111°, the density of the monomer and the density of the polymer were the
same.

$$II \xrightarrow[\substack{\text{di-}\underline{\text{tert}}\text{-butyl} \\ \text{peroxide} \\ 43\%}]{130°} \left[CH_2-\underset{\underset{CH_2}{\|}}{C}-CH_2-O-\underset{\underset{O}{\|}}{C}-O-CH_2-CH_2-CH_2 \right]_x \quad [\eta]^{25°}_{CHCl_3} = 0.11$$

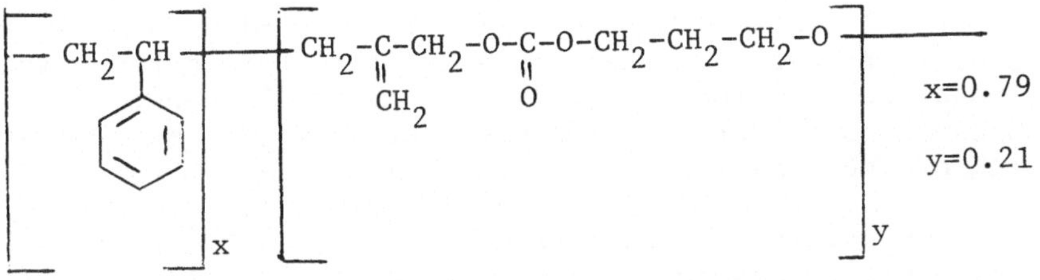

$$\left[\begin{array}{c} CH_2-CH \\ \bigcirc \end{array}\right]_x \left[\begin{array}{c} CH_2-C-CH_2-O-C-O-CH_2-CH_2-CH_2-O \\ \parallel \quad\quad\quad \parallel \\ CH_2 \quad\quad\quad O \end{array}\right]_y$$

x=0.79

y=0.21

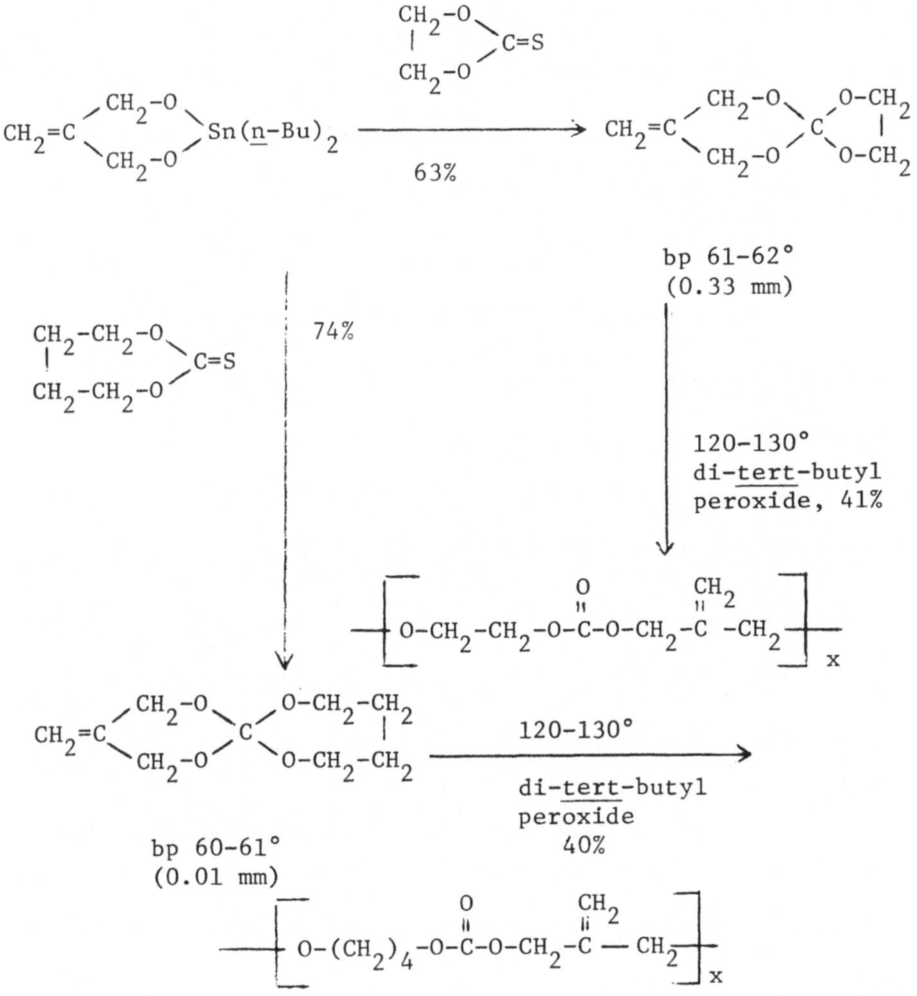

63%

bp 61–62°
(0.33 mm)

74%

120–130°
di-tert-butyl
peroxide, 41%

$$\left[\begin{array}{c} \quad\quad\quad O \quad\quad CH_2 \\ \quad\quad\quad \parallel \quad\quad \parallel \\ O-CH_2-CH_2-O-C-O-CH_2-C-CH_2 \end{array}\right]_x$$

120–130°

di-tert-butyl
peroxide
40%

bp 60–61°
(0.01 mm)

$$\left[\begin{array}{c} \quad\quad O \quad\quad CH_2 \\ \quad\quad \parallel \quad\quad \parallel \\ O-(CH_2)_4-O-C-O-CH_2-C — CH_2 \end{array}\right]_x$$

When the 3-methylene derivative was mixed with an equal
amount of styrene in the presence of di-tert-butyl peroxide
   and the reaction was stopped at below 30% conversion, a
soluble copolymer was obtained containing 79% styrene and
21% of the linear polycarbonate units.

   By a very similar synthetic scheme, other unsaturated
spiro ortho carbonates were prepared. Bulk polymerization
or solution polymerization in chlorobenzene gave soluble
polymer if the reaction was stopped at low conversion.
Both monomers gave cross-linked resins at high conversion.

Since the ethylene ketene acetal underwent ring-opening polymerization quite readily, it appeared desirable to prepare a related spiro ortho ester for use as a monomer with either no change in volume or slight expansion. Thus an unsaturated spiro ortho ester was synthesized by the accompanying reactions. Polymerization of the 2-methylene-1,4,6-trioxaspiro[4.4]nonane with benzoyl peroxide produced a polymer containing keto groups and ester groups with essentially no change in volume. Apparently the radical adds to the double bond to produce the stabilized intermediate radical which can undergo ring opening with the formation of the keto group. Although the acetal radical can theoretically cleave in two directions, the main reaction involves a second ring opening. This approach should allow the synthesis of a wide variety of unsaturated spiro ortho esters containing larger size rings which will produce large increases in volume upon polymerization.

It should be pointed out that not all unsaturated oxygen-heterocyclic monomers undergo ring-opening polymerization. The ketal of cyclohexanone with 2-hydroxymethylallyl alcohol when copolymerized with vinyl acetate at 130° produced a copolymer in which the ring did not open. It may be that at a higher temperature or with a more sluggish comonomer some ring opening will occur.

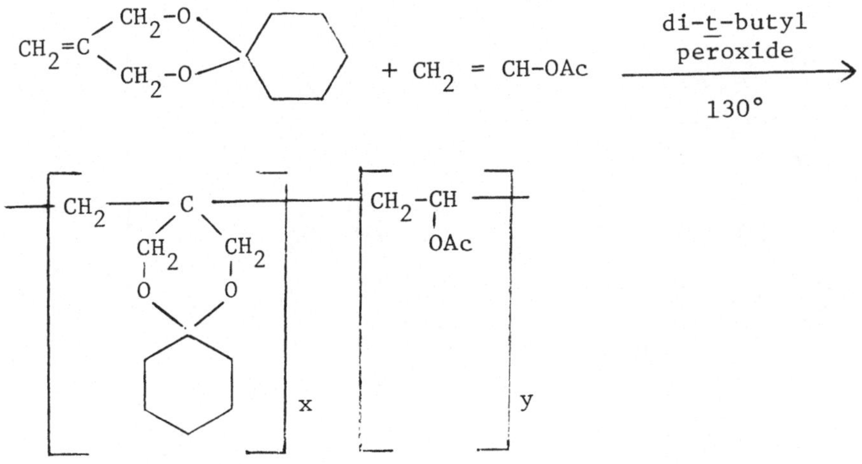

In the previous examples discussed the three-membered
rings had sufficient ring strain to undergo ring opening
even in carbocyclic systems but a five- or six-membered
ring would open under radical conditions only if some other
driving force was present to promote the ring-opening.
Since four-membered rings represent systems in between these
two extremes, they appeared to be interesting cases for
study. The ring opening of cyclobutylcarbinyl is exothermic
[$\Delta H°$ = -4.0 kcal/mol compared to a $\Delta H°$ of -5.1 kcal/mol for
the ring opening of cyclopropylcarbinyl] and the E has been
estimated at 13.0 kcal/mol corresponding to a k of about
4.5×10^3 s^{-1} at 25° C, which are less than the values listed
in Table I for cyclopropylcarbinyl ring opening. Finally
cyclobutylcarbinyl radicals can be observed in ESR at temperatures[19] at which cyclopropylcarbinyl radicals undergo
ring opening. Studies with β-pinene show that ring-opening
can compete with relatively slow abstraction reactions but
can not compete with relatively fast abstraction reactions.
For example, β-pinene will react with carbon tetrachloride
to give a ring-opened product[20] but with thioacetic acid
to give the nonring-opened product.[21]

It was reasoned that if β-pinene was copolymerized with a monomer in which the addition step was relatively slow and involved some steric hindrance that there was a good chance that a copolymer containing ring-opened units could be obtained. Thus maleic anhydride appeared to be an attractive choice since it did not have a high tendency to homopolymerize and was bulky enough to give considerable steric hindrance in the addition step. When equivalent quantities of β-pinene and maleic anhydride were heated at 130° in the presences of di-t-butyl peroxide, an essentially regular alternating copolymer containing the ring-opened β-pinene units was obtained.

Although the oxetane ring possesses considerable ring strain, the corresponding radical does appear to open at a rapid rate. For example, 2-oxetanyl gives no signals in the ESR spectrum for the ring-opened product, which is in contrast to the five- and six-membered acetals which give the ring-opened radical signals at room temperature.[22] However, when 2-methyleneoxetane[23] was copolymerized with maleic anhydride at 130° C, a copolymer was obtained containing the

$$
\begin{array}{c}
\underset{\underset{\text{O-CH}_2}{|\quad|}}{\text{CH}_2\text{=C-CH}_2} \;+\; \underset{\underset{\text{O}^{\diagdown}\text{O}^{\diagup}\text{C}^{\diagdown}\text{O}}{||\quad||}}{\text{CH = CH}} \xrightarrow{130^\circ}
\left[\text{CH}_2\text{-}\underset{\text{O}}{\overset{||}{\text{C}}}\text{-CH}_2\text{-CH}_2\right]_x
\left[\underset{\text{O}^{\diagdown}\text{O}^{\diagup}\text{C}^{\diagdown}\text{O}}{\text{CH --- CH}}\right]_y
\end{array}
$$

$\downarrow$ R° $\qquad\qquad\qquad\qquad\qquad\qquad\uparrow$ repeat

$$
\underset{\underset{\text{O-CH}_2}{|\quad|}}{\text{R-CH}_2\text{-}\overset{\circ}{\text{C}}\text{-CH}_2}
\longrightarrow
\underset{\underset{\text{O CH}_2^\circ}{|\quad|}}{\text{R-CH}_2\text{-C-CH}_2}
\longrightarrow
\text{R-CH}_2\text{-}\underset{\text{O}}{\overset{||}{\text{C}}}\text{-CH}_2\text{-CH}_2\text{-CH --- CH}^c \;\;\underset{\text{O}^{\diagdown}\text{O}^{\diagup}\text{C}^{\diagdown}\text{O}}{}
$$

ring-opened units. Apparently the addition of the tertiary
2-oxetanyl radical to itself or to maleic anhydride is slow
enough because of steric factors to allow the ring opening
to compete. The copolymerization of this monomer appears
to be another way to introduce a carbonyl into the backbone
of an addition polymer to render it possibly photodegradable.

A commercially available monomer that is structurally
related to 2-methylene-oxetane is ketene dimer. However,
almost nothing has been reported concerning its polymerization by free radical procedures. When ketene dimer was
heated with di-t-butyl peroxide at 130° in a sealed tube,
a small pressure built up and the resulting polymer con-
sisted of mostly 1,3-diketone units. When the polymerization was carried out in a solvent at atmospheric pressure,
carbon monoxide was liberated and a copolymer consisting
of mainly monoketone units was obtained.

Apparently the intermediate radical undergoes ring
opening to give the carbonyl radical. In the sealed tube
only a little decarbonylation takes place and the copolymer
contains mostly the β-diketone structure. When the reaction
is carried out at atmospheric pressure, much more decarbonylation takes place and the copolymer consists of mostly
the simple ketone units. Since ketene dimer is an article
of commerce, this monomer appears to be an excellent starting material for the introduction of keto groups into an
addition polymer.

$$CH_2=C-CH_2 \quad \xrightarrow[130°]{R°} \quad R-CH_2-\overset{\overset{\text{o}}{\|}}{C}-CH_2$$
$$\overset{|}{O}-\overset{|}{C}=O \qquad\qquad\qquad O--C=O$$

$$R-CH_2-\overset{\overset{\text{o}}{}}{\underset{\|}{C}}-CH_2° \quad \xleftarrow{\quad -CO \quad} \quad R-CH_2-\overset{|}{C}-CH_2$$
$$\qquad\qquad O \qquad\qquad\qquad\qquad O \quad \underset{\text{o}}{C}=O$$

open to air ↘   ↙ sealed tube

$$\cdots \cdot CH_2-\overset{\overset{\text{O}}{\|}}{C}-CH_2 \overbrace{}^{x} \quad CH_2-\overset{\overset{\text{O}}{\|}}{C}-CH_2-\overset{\overset{\text{O}}{\|}}{C} \overbrace{}^{y}$$

REFERENCES

1. Presented at the 9th Biennial Polymer Symposium, Key Biscayne, Florida, November 20, 1978.

2. Supported in part by grants from the Naval Air Systems Command, the National Institute of Dental Research and the National Science Foundation.

3. G. B. Butler and R. J. Angelo, J. Am. Chem. Soc., 79, 3128 (1957).

4. T. Takahashi, J. Polym. Sci., A6, 403 (1968).

5. I. Cho and K. D. Ahn, J. Polym. Sci., Polymer Letters, 15, 751 (1977).

6. L. A. Errede, J. Polym. Sci., 49, 253 (1961).

7. B. Maillard, D. Forrest, and K. U. Ingold, J. Am. Chem. Soc., 98, 7024 (1976).

8. B. Maillard, M. Cazaux, and R. Lalande, Bull. Soc. Chim. Fr., 1973, 1368.

9. F. Beyerstedt and S. M. McElvain, J. Am. Chem. Soc., 58, 529 (1936).

10. W. J. Bailey and E. A. Ofstead, unpublished data.

11. W. J. Bailey, R. L. Sun, H. Katsuki, T. Endo, H. Iwama, R. Tsushima, K. Saigo, and M. M. Bitritto, "ACS Symposium, No. 59, Ring-Opening Polymerization", T. Saegusa and E. Goethals, Eds., American Chemical Society, Washington, D.C., 1977, p. 38.

12. W. J. Bailey, H. Katsuki, and T. Endo, Amer. Chem. Soc., Div. Polym. Chem., Prepr., 15, 445 (1974).

13. T. Endo and W. J. Bailey, J. Polym. Sci., Polym. Letters Ed., 13, 193 (1975).

14. V. P. Thompson, E. F. Williams, and W. J. Bailey, J. Am. Dental Assoc., In press.

15. W. J. Bailey and T. Endo, J. Polym. Sci.: Polym. Sym. 64, 17 (1978).

16. T. Endo and W. J. Bailey, Makromol. Chem., 176, 2897 (1975).

17. T. Endo and W. J. Bailey, J. Polym. Sci., Polym. Chem. Ed., 13, 2525 (1975).

18. W. J. Bailey and T. Endo, J. Polym. Sci., Polym. Chem. Ed., 14, 1735 (1976).

19. P. M. Blum, A. G. Davies, and R. A. Henderson, J. Chem. Soc., Chem. Comm., 569 (1978); E. A. Hill, A. T. Chen, and A. Doughty, J. Am. Chem. Soc., 98, 167 (1976).

20. G. du Pont, R. Dulov, and G. Clement, Bull. Soc. Chim. Fr., 1950, 1056.

21. F. G. Bordwell and W. A. Hewett, J. Am. Chem. Soc., 79, 3493 (1957).

22. A. J. Dobbs, B. C. Gilbert, and R. O. C. Norman, J.
     Chem. Soc., A, 124 (1971); J. Chem. Soc., Perkin Trans.,
     2, 786 (1972).

23. P. Hudrlik, A. Hudrlik, and C. Wan, J. Org. Chem.,
     40, 1116 (1975).

HIGH POLYMERIC ORGANOPHOSPHAZENES - MACROMOLECULES WITH A DIFFERENCE

Harry R. Allcock

Department of Chemistry
The Pennsylvania State University
University Park, Pennsylvania 16802

The discovery and development of a new polymer system is nearly always a stimulating event in macromolecular science. It opens up novel avenues for fundamental studies, yields additional opportunities to test established or developing theories and, at the same time, provides unusual materials for technology.

Our research over the past several years has been devoted to the exploratory synthesis and characterization of an entirely new class of polymers - the poly(organophosphazenes). These polymers are not only new, but they possess a highly unusual backbone structure composed of an __inorganic__ chain of alternating phosphorus and nitrogen atoms. Some typical poly(organophosphazene) structures are shown in I-IV.

$$
\left[N = P \begin{matrix} OR \\ | \\ | \\ OR \end{matrix} \right]_n \qquad
\left[N = P \begin{matrix} NHR \\ | \\ | \\ NHR \end{matrix} \right]_n \qquad
\left[N = P \begin{matrix} NR_2 \\ | \\ | \\ NR_2 \end{matrix} \right]_n \qquad
\left[N = P \begin{matrix} R \\ | \\ | \\ R \end{matrix} \right]_n
$$

I   II   III   IV

In these macromolecules the group R is an organic residue. Hence, an appropriate general name for these compounds is that of "inorganic-organic" or "heteroatom" polymers.

In this paper I will attempt to point out the ways in which these polymers are <u>different</u> from conventional organic macromolecules.   I shall also discuss how these differences can be and are being exploited both in fundamental research and in the development of new materials for technology.   Finally, I will suggest some of the structural reasons why phosphazene polymers have properties that are unusual and unexpected.

GENERAL METHOD OF SYNTHESIS

The most striking difference between conventional polymers and poly(organophosphazenes) is in their method of synthesis.   The normal techniques for the synthesis of macromolecules – i.e. the polymerization of unsaturated monomers or the condensation reactions of difunctional monomeric reagents – are not applicable to polyphosphazene synthesis.   Monomers of structure, $N\equiv P(OR)_2$, $N\equiv P(NHR)_2$, $N\equiv P(NR_2)_2$, or $N\equiv PR_2$, have not yet been isolated. Direct ring-opening polymerization cannot yet be used for the synthesis of <u>organo</u>phosphazene high polymers.   Even though the first step in the synthesis procedure is a ring-opening polymerization of a cyclic halogeno-phosphazene (see below), cyclic <u>organo</u>phosphazenes usually resist polymerization.

The key to the synthesis of poly(organophosphazenes) is the use of a preformed, linear, high polymeric halogenophosphazene as a <u>highly reactive intermediate</u> for substitution reactions.   A few organic polymers are prepared by the modification of preformed macromolecules (for instance, the formation of poly(vinyl alcohol) from poly(vinyl acetate), or the chloromethylation of poly-styrene), but this method of synthesis cannot be applied generally because of the low reactivity of most organic polymers and the well-known problems that result from chain-coiling in solution or from the deactivation induced by charge generation on nearby repeating units.

The overall synthesis routes for poly(organo-phosphazenes) are shown in Scheme 1.

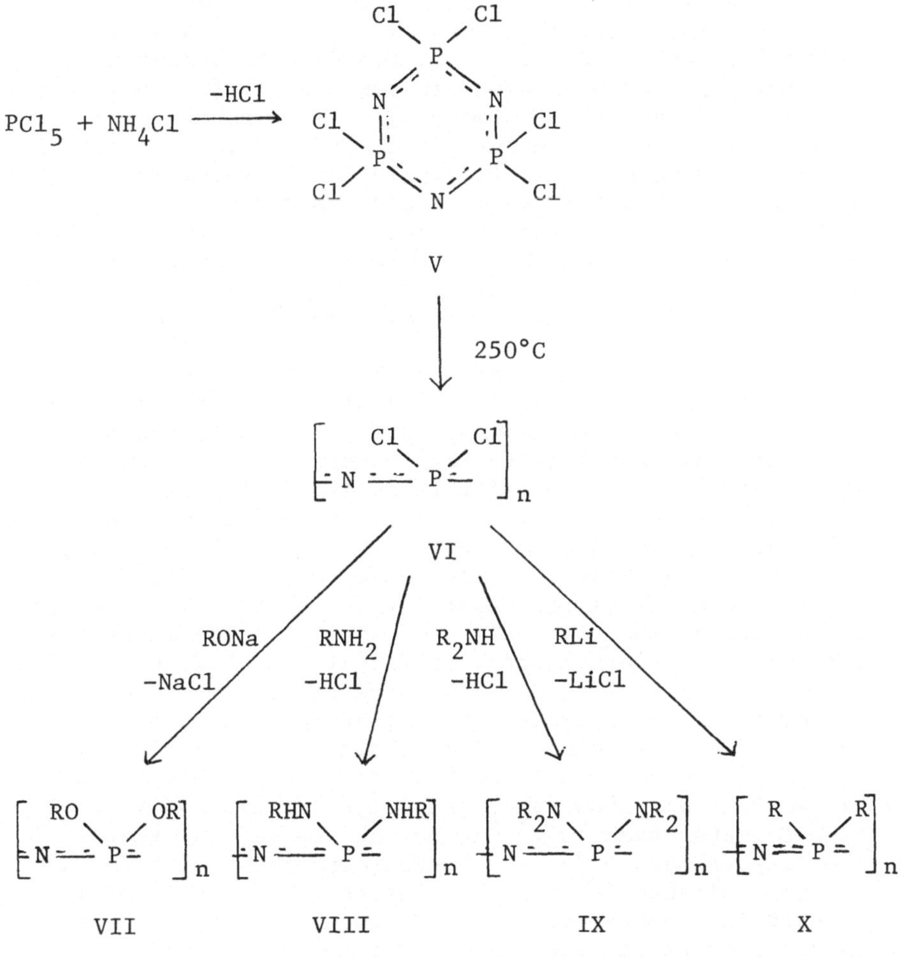

PCl5 + NH4Cl
-HCl
Cl
P
N
V
250°C
VI
RONa
-NaCl
RNH2
-HCl
R2NH
-HCl
RLi
-LiCl
RO
OR
RHN
NHR
R2N
NR2
R
n
VII
VIII
IX
X

$n \simeq 15{,}000$

Scheme 1. General synthesis routes to poly-(organophosphazenes)

The formation of hexachlorocyclotriphosphazene (V) from
phosphorus pentachloride and ammonium chloride or ammonia
has been known since the work of Liebig and Wöhler in
1834.[1]   Similarly, the thermal polymerization of V to a
rubbery, crosslinked form of poly(dichlorophosphazene)
(VI) was reported by Stokes[2] as early as 1897.   However,
for over 70 years this polymer was viewed merely as a
laboratory curiosity because it is hydrolytically unstable
in the atmosphere and is insoluble in all solvents.   Hence,
the early work on this polymer was restricted to fundamental studies of the solid state properties, such as its
elastic behavior and X-ray diffraction characteristics.[3]

My own interest in this field arose from the
recognition (based on reaction mechanistic arguments)
that the phosphorus-nitrogen backbone itself should be
hydrolytically stable, and that replacement of the halogen
atoms by organic residues should yield substituted macromolecules that are hydrolytically stable.   These ideas
were tested first with small-molecule cyclic chloro-
phosphazenes, such as V.[4]   But the extension of these
principles to the macromolecular analogue (VI) proved to
be a serious problem.   An insoluble polymer is an
unsuitable material for substitution reactions.   However,
during the course of a series of kinetic experiments on
the thermal conversion of V to VI, it was observed that
the "swellability" of VI in benzene or tetrahydrofuran
decreased as the polymerization time was increased.[5]
Hence, it was reasoned that at short polymerization times
a soluble and uncrosslinked form of VI might be formed.
In fact, it was shown that the polymerization of V to VI
is a two-step reaction.[6-8]   During the initial stages of
the polymerization (up to ∿70-75% conversion of V to VI)
an uncrosslinked form of VI is formed.   This polymer is
soluble in a number of organic solvents, such as benzene,
toluene, or tetrahydrofuran.   Beyond this stage, the
polymer crosslinks rapidly.   The mechanism of this
crosslinking process is still not fully understood,
although traces of water will accelerate the process,
possibly by yielding P-O-P bridging links.[9]

In solution, the uncrosslinked form of VI is a highly
reactive species.   It reacts rapidly with alkoxides,
aryloxides, amines, and some organometallic reagents to

yield polymers such as I-IV.[6-8]   This high nucleophilic
reactivity is surprising, even when viewed against a
background of the known reactions of small-molecule
phosphorus-halogen compounds, such as PCl₃, PCl₅,
(NPCl₂)₃ (V), etc..[10]   It suggests that the coiling of
the macromolecule in solution does not necessarily lead to
a steric retardation of substitutive processes.   In some
cases - for example, in the reaction of VI with a large
excess of sodium trifluoroethoxide - the total substitution
process appears to be almost instantaneous.[7]

    We have recently developed a modification to this
general synthesis route, specifically for the purpose of
preparing polymers of structure, IV.   Poly(dichlorophosphazene) (VI) reacts with organometallic species
such as Grignard or organolithium reagents by two different
reaction pathways - one favorable and one distinctly
unfavorable.   These two reactions are alkylation or
arylation (XII) on the one hand, and chain-cleavage (XIII)
on the other.[11,12]

Because the chain cleavage reaction is presumably favored
by a high electron-density in the lone-pair-electron
orbital at skeletal nitrogen, we have assumed that the use
of the more electronegative fluorine atoms in poly-
(difluorophosphazene) may favor halogen substitution at the
expense of chain cleavage.   Poly(difluorophosphazene)
(XV) can be prepared by the high pressure, high temperature
polymerization of hexafluorocyclotriphosphazene (XIV).[13,14]
Once again this is a two-step process process.   In the
first step the reaction mixture contains only a decreasing
amount of XIV and an increasing proportion of uncrosslinked XV.   In the second stage, XV crosslinks, often
when the conversion of XIV to polymer has risen above
∿70%.[13,14]   The reactions of XV with organometallic
reagents yield alkylated or arylated high molecular
weight polymers, although 100% alkylation or arylation
has not yet been achieved without appreciable chain
cleavage.

XIV   XV

VERSATILITY

In conventional polymer research, the synthesis of a
new macromolecule usually involves the preparation or
acquisition of a new monomer or combination of monomers,
followed by attempts to discover suitable, individual sets
of reaction conditions that will permit the conversion of
that monomer to a high polymer.   By contrast, in polyphosphazene chemistry an enormous range of different
polymers can be prepared by relatively simple techniques
from one or two preformed polymeric starting materials.

This means that the polymerization problem is a relatively trivial aspect of the synthesis. It also means that <u>high</u> polymers can be prepared by straightforward procedures, and that any differences in chain length can be attributed to chain-cleavage side reactions rather than to the usual chain-transfer or termination problems inherent in the traditional methods of polymer synthesis. In this sense, polyphosphazene chemistry represents a new and rather unusual approach to both fundamental and applied polymer chemistry. Thus, close comparisons can be made between polymers that have different side groups or mixtures of side groups but virtually identical chain lengths and molecular weight distributions. The value of this feature for both fundamental structure-property studies and for technology cannot be overemphasized.

How many different poly(organophosphazenes) are accessible by this synthesis route? Taking into account the available alcohols, phenols, primary and secondary amines, and organometallic reagents, it seems clear that the phosphazene homopolymers accessible by known synthesis routes rival in numbers all the known organic polymers. When the permutations of mixtures of different substituent groups attached to the chain are taken into account, the number of possible polymers is almost beyond calculation. Different substituent groups impart different physical and chemical properties. Thus, it will be clear that the polyphosphazene system is a major new area of polymer chemistry that rivals or surpasses the more traditional macromolecular systems in both scope and versatility. At the present time, roughly 80 different poly(organophosphazenes) have been prepared and characterized.[6-8,13,14,15-27] Clearly, the opportunities for further research and technological development in this area are extremely promising.

LIMITS AND RESTRICTIONS

This unusual synthetic versatility can, in principle, give rise to an almost unprecedented range of new macromolecules. However, it is important to note that certain restrictions exist with respect to the types and combinations of different substituent groups that can be attached to the polyphosphazene chain.

First, the nucleophilic substitution reactions of VI
generally fall into the category of S_N2-type replacements.
Hence, they are affected by the nucleophilicity and steric
characteristics of the attacking nucleophile and by the
leaving-group ability of the halogen.   Second,
restrictions exist when a prospective nucleophile
possesses two or more potential nucleophilic sites.
For example, a difunctional reagent (a diamine or diol)
could crosslink the chains.   Third, as mentioned
previously, the possibility exists that the cleavage of
phosphorus-nitrogen skeletal bonds might become
competitive with phosphorus-halogen bond cleavage. ˙ A few
examples will illustrate some of the specific restrictions
that have been identified.

The reactions of amines with poly(dihalophosphazenes)
are, in general, more sensitive to mechanistic restrictions
than are the substitutions by alkoxides or aryloxides.
For example, diethylamine replaces only one chlorine per
phosphorus in VI to yield polymers of structure, XVI.[15]

$$\left[\begin{array}{c} Cl \\ | \\ -N = P - \\ | \\ Cl \end{array}\right]_n \quad \xrightarrow[-HCl]{Et_2NH} \quad \left[\begin{array}{c} NEt_2 \\ | \\ -N = P - \\ | \\ Cl \end{array}\right]_n$$

VI   XVI

Diphenylamine apparently undergoes no substitution at all.
These results reflect the sensitivity of the aminolysis
reaction to steric effects and to the nucleophilicity of
the amine.   Moreover, if poly(difluorophosphazene) (XV)
is used as a polymeric intermediate, even primary amines
replace only one fluorine per phosphorus,[14] under
conditions where total halogen replacement occurs with
poly(dichlorophosphazene).   This effect is ascribed
partly to the poor leaving-group ability of fluorine
compared to chlorine.   Steric effects are particularly

$$\left[-N \doteq P \doteq \begin{matrix} F \\ | \\ | \\ F \end{matrix} \right]_n \quad \xrightarrow[-HF]{RNH_2} \quad \left[-N \doteq P \doteq \begin{matrix} NHR \\ | \\ | \\ F \end{matrix} \right]_n$$

XV   XVII

noticeable when bulky nucleophiles such as the azo-dye
depicted in XVIII[17] or the steroidal anion shown in XIX
are employed.[28]   Only one of these molecules can be
introduced every three or four repeating units along the
polymer chain, and some difficulty is encountered when
attempts are made to replace the remaining halogen atoms
by less hindered nucleophiles.

XVIII   XIX

     The crosslinking reactions by difunctional reagents
are facile processes.   Aliphatic or aromatic diamines or
the alkoxides generated from diols readily crosslink the
chains, either by halogen replacement or, in some cases,
by the displacement of organic groups already present.[29]
Even ammonia or methylamine can function as crosslinkage
agents.   However, methylamine does not crosslink the
chains at low temperatures, and ethylamine and higher
alkyl or primary amines function exclusively as mono-
rather than di-nucleophiles.

Perhaps the most serious restriction to the diversif-
ication of polyphosphazene structures is found in the
tendency of many reagents to induce chain cleavage.   The
role of organometallic reagents in chain cleavage has
already been mentioned.   However, carboxylic acids and
their alkali metal salts are particularly effective
chain-cleavage agents.   The mechanisms of these cleavage
reactions are only partly understood.   Nevertheless,
this reaction pathway precludes the use of many
biologically active agents as substituent groups.

DIVERSITY OF CHEMICAL PROPERTIES

The unique feature of the polyphosphazene system is
the ease with which the side group modifications can be
introduced into the macromolecules.   These modifications
result in changes in both the chemical and physical
properties.   The chemical properties will be mentioned
here, and the physical characteristics will be discussed
in a later section.

The chemical characteristics of poly(organophosphazenes)
can be understood in terms of two factors - the nature of
the backbone and the structure of the side group.   The
chemistry of the backbone is dominated by the presence of
the lone-pair electrons on the skeletal nitrogen atoms.
The basicity of these nitrogen atoms facilitates pro-
tonation, coordination to metals, or hydrogen bonding to
water or other protic solvents.   For example, the polymer
$[NP(NHCH_3)_2]_n$ forms acid-base "salts" with hydrohalides,
functions as a polymeric ligand for transition metals
such as platinum,[30] and at the same time is soluble in
water or alcohols.

An equally powerful influence on the chemical
properties is exerted by the side group structure -
sometimes in opposition to the skeletal influence.   For
example, although the CH_3NH- side group confers water
-solubility on the polymer, fluorinated side groups, such
as CF_3CH_2O- or $CF_3CF_2CH_2O-$, give rise to hydrophobicity
and water-insolubility.   However, these latter side
groups provide solubility in ketones or fluorocarbons.
The phenoxy group imparts solubility in hot, aromatic
hydrocarbons, but insolubility in nearly all other media.

Thus the hydrophobicity or hydrophilicity of a polymer
can be varied over a wide range by a choice of suitable
side groups.

The hydrolytic stability of a polyphosphazene is
markedly dependent on the type of side group.   Nearly all
poly(organophosphazenes) are stable to aqueous media, but
the most hydrophobic species are remarkably resistant to
hydrolytic degradation.   The polymers, $[NP(OCH_2CF_3)_2]_n$
and $[NP(OC_6H_5)_2]_n$, are unaffected after years of immersion
in strong aqueous sodium hydroxide solution.   However, a
limited number of side groups are hydrolytic destabilizing
groups.   For example, polymers that possess $-NH_2$ or
$-NHCH_2COOR$ groups hydrolyze slowly in contact with
moisture.[16]

The thermal stability, also.depends on the types of
side groups present.   Amino side groups usually condense
at high temperatures to yield crosslinked matrices.
Alkoxy- or aryloxyphosphazene high polymers (where
OR = OCH_2CF_3 or OC_6H_5) depolymerize to cyclic oligomers
at temperatures above 200-250°C.   This behavior is
reminiscent of the "reversion" reactions that are
characteristic of poly(organosiloxanes).

Finally, the choice of a suitable side group can
modify the metal-adduct coordination properties of the
polymer (imidazoyl groups),[31] or the optical absorption
properties (dyestuffs,[17] aromatic amino residues, ketonic
groups, etc.).   In all these characteristics, polyphosphazenes offer the prospect of new combinations of
properties that are difficult or impossible to accomplish
by the homopolymerization or copolymerization of organic
monomers.

RELATIONSHIP OF PROPERTIES TO USES

The technological side of polyphosphazene chemistry
is still in its infancy.   Even so, it is possible to
recognize the emergence of four lines of technology where
polyphosphazenes may have a significant impact.   These
are the areas of elastomer technology, films and surface
coatings, fibers, and biomedical research and engineering.
These will be discussed briefly in turn.

A need exists for elastomeric materials that are
resistant to oils, greases, or solvents, that have low
glass transition temperatures, do not burn, and can
withstand moderate to high temperatures, ozonolysis, or
ultraviolet degradation.   Such materials would be used
in automotive or aerospace applications as O-rings,
gaskets, pipes, or energy absorbing devices or in
non-burning thermal or electrical insulation.   Poly-
(organophosphazenes) that have two or more substituent
groups attached to the chain[18] have been developed
commercially[32] for high-technology elastomer
applications.[19,20,33]   In this respect they appear to be
superior to silicone rubber or organic high-performance
elastomers for many uses.

Certain advantages can be foreseen for the use of
polyphosphazenes as films or surface coatings.   The
surface properties of the polymers can be modified to a
subtle degree.   Moreover, the ultraviolet stability of
polymers such as $[NP(OCH_2CF_3)_2]_n$ is especially high.[34]
Those photolytic reactions that do occur appear to involve
reactions of the side groups rather than the backbone.
The phosphorus-nitrogen bonds of the skeleton are far less
susceptible to free-radical induced chain cleavage than are
the carbon-carbon or carbon-oxygen bonds of conventional
macromolecules.   The flame-resistance of most polyphosphazenes is an added advantage.

A number of microcrystalline homopolymers, such as
$[NP(OCH_2CF_3)_2]_n$ or $[NP(OC_6H_5)_2]_n$, can be solution
-fabricated into fibers.   These do not burn.   However,
the higher cost of most poly(organophosphazenes) compared
to the well-known organic fiber-forming polymers may
restrict their use to specialized, high-performance
applications.

Polyphosphazenes may provide particular advantages
over their organic counterparts in the field of biomedical
applications.   For artificial organ research, materials
can be synthesized that have specific surface properties,
extreme stability under hydrolytic or oxidative conditions,
and minimal interactions with blood or living tissues.
Polymers that possess fluoroalkoxy or aryloxy side groups

have shown promise in animal implantation tests.[35]   In
general, this type of application involves the utilization
of polymers that can be prepared by existing techniques.
More intriguing is the prospect that new types of polyphosphazenes can be employed as membrane materials, as
polymer-bound dyes or sunscreen agents, or as carrier
molecules for the slow release of chemotherapeutic drugs.
For example, amino acid ester derivatives of formula XX
have been prepared.[16]

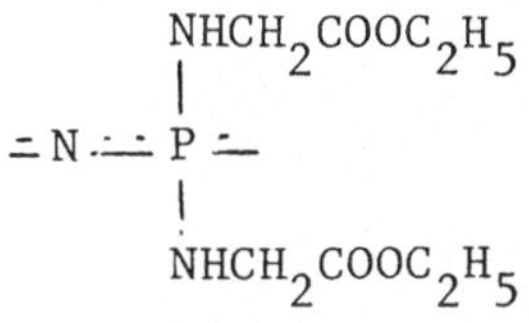

$$=N-P-$$

with substituents $NHCH_2COOC_2H_5$ (top) and $NHCH_2COOC_2H_5$ (bottom)

XX

$$=N=P-$$

with substituents O—steroid (top, CH_3) and NHR (bottom)

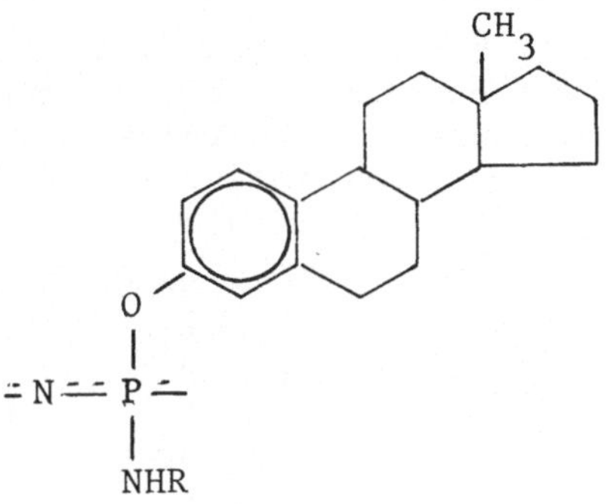

XXI

Unlike most polymers (except poly(amino acids),poly(lactic
acid), or poly(glycolic acid)), compounds such as XX
hydrolyze slowly to yield ethanol, amino acid, phosphate,
and ammonia.   Thus, the polymer is biocompatible in the
sense that it decomposes to materials that can be
metabolized or excreted.   Recent work in our laboratory
has exploited the linkage of steroid molecules (XXI) to
a polyphosphazene skeleton[28] with a view to the utilization
of these compounds as slow-release drugs.   Furthermore,
we have recently examined the use of poly[bis(methylamino)-
phosphazene], $[NP(NHCH_3)_2]_n$, as a coordinative ligand for
platinum anticancer drugs.[30]   The $PtCl_2$ unit is bound to
the skeletal nitrogen atoms to form a square planar
arrangement (XXII).

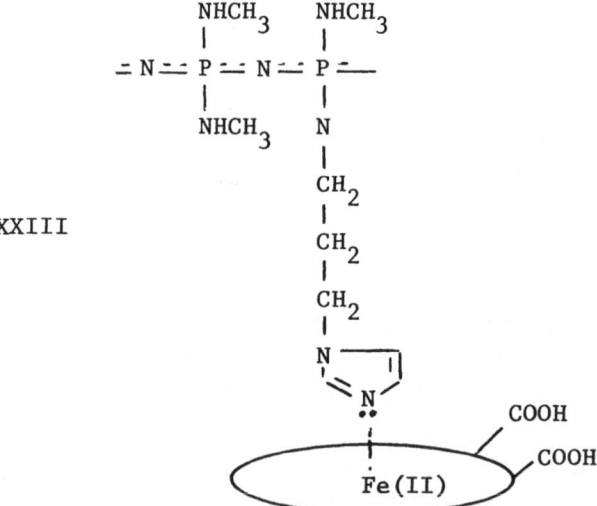

XXII

Models for hemoglobin and myoglobin have also been prepared
in which water-soluble polyphosphazenes replace the protein
components, and coordinate to heme.   One such structure
is shown diagramatically in XXIII.31

XXIII

Given the synthetic versatility in this polymer system,
it is clear that a great many chemotherapeutic or
biologically interesting molecules could be bound to
phosphazenes.   Clearly the pace of new biomedical
developments is critically dependent on the progress of the
fundamental synthetic research.

RELATIONSHIP OF STRUCTURE TO PHYSICAL PROPERTIES

The physical and morphological properties of polyphosphazenes are unusual.   Polymers are known that cover
the range from low temperature amorphous elastomers to high
melting crystalline glasses.   Many polyphosphazenes are
microcrystalline, flexible, film- or fiber-forming materials.
Of course, these characteristics can be found in any
number of organic polymers.   However, the unique feature
of polyphosphazenes is the manner in which the morphological
characteristics can be changed by subtle alterations in the
side group structure.   Furthermore, certain polyphosphazenes have some of the lowest glass transition
temperatures that have been observed in any class of polymers
(in the range of -95°C to -60°C).   The fact that polymers
such as $(NPCl_2)_n$, $(NPF_2)_n$, $[NP(OCH_3)_2]_n$, $[NP(OCH_2H_5)_2]_n$,[7]
or species of formula $[NP(OR)(OR')]_n$[18,19,22] are elastomers
at room temperature and below is surprising in view of the
inorganic character of the backbone.

It is generally agreed that the physical property
differences between organic-type polymers can be
attributed to differences in crystallinity, to side group
steric or polar effects, and occasionally to the existence
of cis-trans isomerism about double bonds in the skeleton. To
what extent are the properties of polyphosphazenes
determined by the same factors?

The preliminary available evidence suggests that only
the first two factors exert an appreciable influence in
the polyphosphazene series.   The multiple bonding in the
phosphazene backbone bears virtually no resemblance to
the unsaturation found in polyacetylene, or in cis- or
trans-1,4-polyisoprene or 1,4-butadiene.   Thus, an
understanding of the differences between polyphosphazenes
and unsaturated organic polymers must begin with a
consideration of the peculiarities of the phosphorus-nitrogen
bond.   The nature of this bond is still a subject for
debate.   The pairing of electrons to form a sigma-bond
framework leaves three electrons on nitrogen and one on
phosphorus still unaccounted for (XXIV).

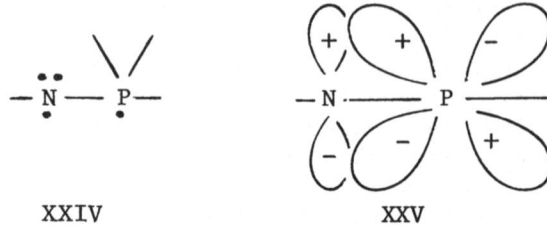

XXIV   XXV

Two of the electrons at each nitrogen atom function as if
they occupy a lone pair (sp^2) orbital, and the basicity
and coordination behavior of the skeleton is compatible
with this interpretation.   The fate of the remaining
electrons – one from phosphorus and one from nitrogen –
is still obscure.   The most attractive explanation is that
the two electrons per repeating unit are accommodated in a
phosphorus 3d orbital and a nitrogen p orbital.   The
overlap of these would then generate a pπ–dπ system
(XXV).[36,37]

    The existence of a pi-electron system of this type
does not imply that similarities are to be expected
between polyphosphazenes and conjugated (or non–conjugated)
organic polymers that possess an unsaturated skeleton.
The symmetry characteristics of d-orbitals are such that
long-range electron-delocalization effects are not
expected and are not observed (most polyphosphazenes are
colorless).   In this respect, polyphosphazenes are unlike
polyacetylene.   Moreover, the normal barriers to internal
rotation that are characteristic of carbon-carbon double
bonds are not detected in polyphosphazenes.   As a P–N
skeletal bond undergoes torsion, the nitrogen p orbital
could, in principle, overlap with any one of the five
phosphorus 3d orbitals.   Thus, in its torsional
characteristics, a sigma –pπ–dπ bond should resemble a
normal organic covalent sigma bond.   Even the bond
lengths are similar.   The carbon–carbon single bond
length is about 1.54–1.56 Å, and the phosphorus–nitrogen
bond distance in cyclic or polymeric phosphazenes is in
the range of 1.52–1.60 Å.

    Strong evidence now exists that the physical properties
of polyphosphazenes are dominated by the size and polarity

of the side groups in a manner that is reminiscent of the
behavior of conventional organic systems.   The difficulty
is to distinguish between the intramolecular influences
of the side group and its effect on the intermolecular
interactions.   Polyphosphazenes possess one advantage
over most synthetic polymer systems in the ease with
which changes in side group structure can be made without
appreciably affecting the chain length or molecular
weight distribution.   Hence, it is instructive to compare
the conformational behavior, glass transition temperatures,
and the tendency for the appearance of crystallinity
and/or elasticity in a closely related set of polymers.
The possibility exists that such comparisons may even be
of value for the interpretation of the properties of
organic macromolecules.

Consider the polymers shown in XV, VI, XXVI-XXIX.

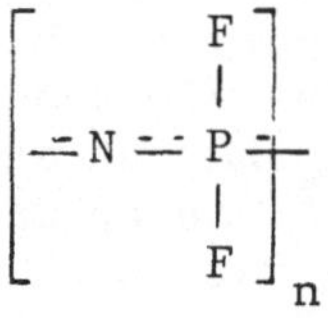

$$
\left[\!-N=P\genfrac{}{}{0pt}{}{F}{F}\right]_n
$$

XV

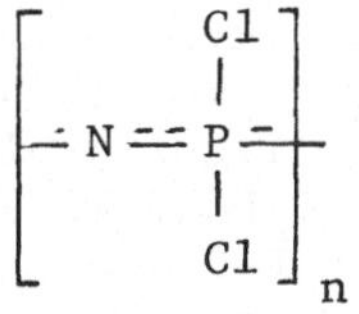

$$
\left[\!-N=P\genfrac{}{}{0pt}{}{Cl}{Cl}\right]_n
$$

VI

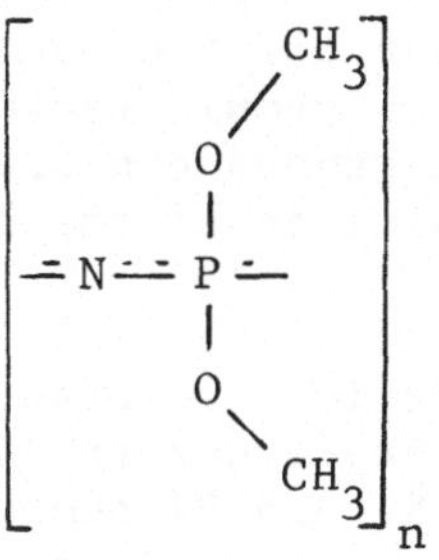

$$
\left[\!-N=P\genfrac{}{}{0pt}{}{O-CH_3}{O-CH_3}\right]_n
$$

XXVI

$$
\left[\!-N=P\genfrac{}{}{0pt}{}{O-CH_2-CH_3}{O-CH_2-CH_3}\right]_n
$$

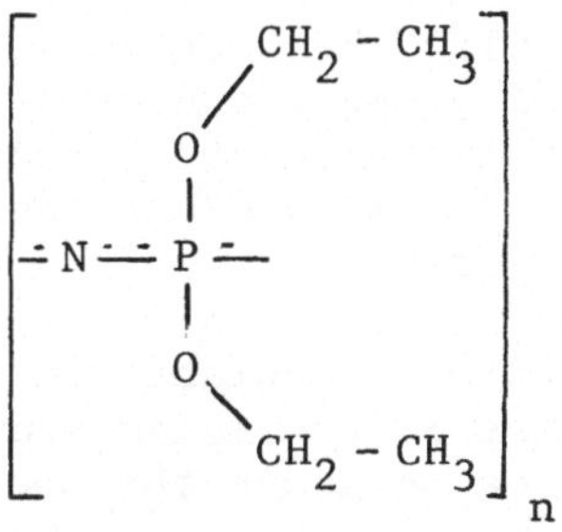

XXVII

$$
\begin{bmatrix}
& & O{-}CH_2 - CF_3 \\
& & | \\
=N{=\!=}P{=\!-} \\
& & | \\
& & O{-}CH_2 - CF_3
\end{bmatrix}_n
\qquad
\begin{bmatrix}
& O{-}\bigcirc \\
& | \\
{-}N{=\!=}P{=\!-} \\
& | \\
& O{-}\bigcirc
\end{bmatrix}_n
$$

XXVIII   XXIX

Any attempt to understand the fundamental influence of the
side group structure must begin with a consideration of
polymers that possess the smallest, least complicated
side groups.   Polymers XV and VI fall into this category.
The more complex polyphosphazenes. such as XXVI – XXIX can
then be visualized as structural "derivatives" of XV, since
the r_o value of oxygen (1.52 Å) is only slightly larger
than that of fluorine (1.47 Å).   The electronegativity
of fluorine is slightly higher than that of oxygen.
However, the P-O bond length (1.58 Å) is appreciably
longer than the P-F length (1.47 Å) but is shorter than
the P-Cl bond distance (1.99 Å).   Thus, a comparison of
the properties of XV or VI with those of XXVI – XXIX
should provide clues to the ways in which the alkyl or
aryl groups in XXVI – XXIX affect the properties.   In
particular, because rotation about the P-O bonds is
possible, the property differences should reflect the
ease with which the various side groups can practice
subtle "avoidance" motions as twisting of the skeleton
takes place.

        First, consider the conformations assumed by these
polymers in the solid state.   Detailed X-ray diffraction
evidence suggests that polymers XV and VI occupy the
conformation shown in XXX.[38,39]   Fiber diagram repeat
distances suggest that XXVIII and XXIX assume the same
conformation.[7]   For XV, this minimum energy conformation
is found only at temperatures below –56°C.   A

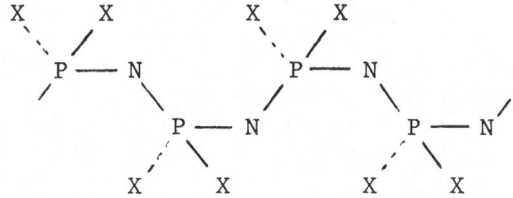

XXX

conformational transition takes place above this
temperature, probably to yield a non-planar helical form.[38]
The existence of this transition at such a low
temperature suggests that the conformation shown in XXX
is not stabilized appreciably by <u>skeletal</u> bonding factors.
Hence, the observation that VI, XXVIII, and XXIX occupy
the same conformation at higher temperatures (∿25°C)
indicates that the side group steric effects are
responsible for this behavior.   Conformational energy
calculations support this view.[40,41]

Second, it is of interest to note the way in which
the glass transition temperatures of these six polymers
vary with the changes in side group structure.   The
lowest value of T_g in $(NPX_2)_n$ is for X = F (-95°C), with
the T_g rising for the other polymers in the order
X = OC_2H_5 (-84°C), OCH_3 (-76°C), OCH_2CF_3 (-66°C), Cl
(-63°C), OC_6H_5 (-8°C).   The introduction of a methoxy
or an ethoxy group in place of fluorine has a surprisingly
small effect on the glass transition temperature and this
is consistent with the idea that the torsional freedom of
the backbone is affected hardly at all by this replacement.
On the other hand, the phenoxy group clearly exerts a
powerful inhibiting influence on the torsional mobility of
the skeleton.

Amino side groups give rise to higher T_g values than
do alkoxy side groups, and this can be attributed to the
enhanced steric effects associated with $-NR_2$ groups, or
the opportunities for hydrogen bonding when $-NHR$ residues
are present.

Crystallinity or the lack of it depends both on the uniformity of the distribution of substituent groups along the chain and on the conformational mobility of the side group.   For example, $[NP(OCH_2CF_3)_2]_n$ and $[NP(OC_6H_5)_2]_n$ are microcrystalline, flexible polymers, but $[NP(OCH_3)_2]_n$, $[NP(OC_2H_5)_2]_n$, and $[NP(OR)(OR')_2]_n$ are non-crystalline elastomers.

Any structural comparisons between polyphosphazenes and organic macromolecules must also take into account (a) the presence of side groups on every other skeletal atom instead of every skeletal atom, (b) the wider bond angle at dicoordinate nitrogen (120-140°) than at tetrahedral carbon or tricoordinate nitrogen, and (c) perhaps the greater variability of the bond angle at dicoordinate nitrogen than at tetrahedral carbon or tricoordinate nitrogen.   In particular, the very low glass transition temperatures found for XV, VI, XXVI, XVII, and XVIII probably reflect the ability of the P-N-P bond to widen considerably as the chain undergoes conformational transitions.   In this respect, polyphosphazenes resemble polysiloxanes.

PROSPECTS FOR THE FUTURE

A number of the synthetic prospects for the field have already been discussed.   Recent work with cyclic phosphazene model compounds has shown that the prospect exists for the modification of side groups already attached to a phosphazene backbone.

For example, the cyclic trimeric compound, XXXI, can be converted to the hexalithio-derivative, XXXII, which in turn reacts with electrophiles, such as carbon dioxide, triphenyltin chloride, or diphenylchlorophosphine, to yield XXXIII, XXXIV, or XXXV.[4]   Compound XXXV functions as a coordinative ligand for transition metals and, hence, the high polymeric analogues are expected to function as carrier molecules for catalysts.

However, much additional synthetic work remains to be done, especially with respect to the reactions of organo-

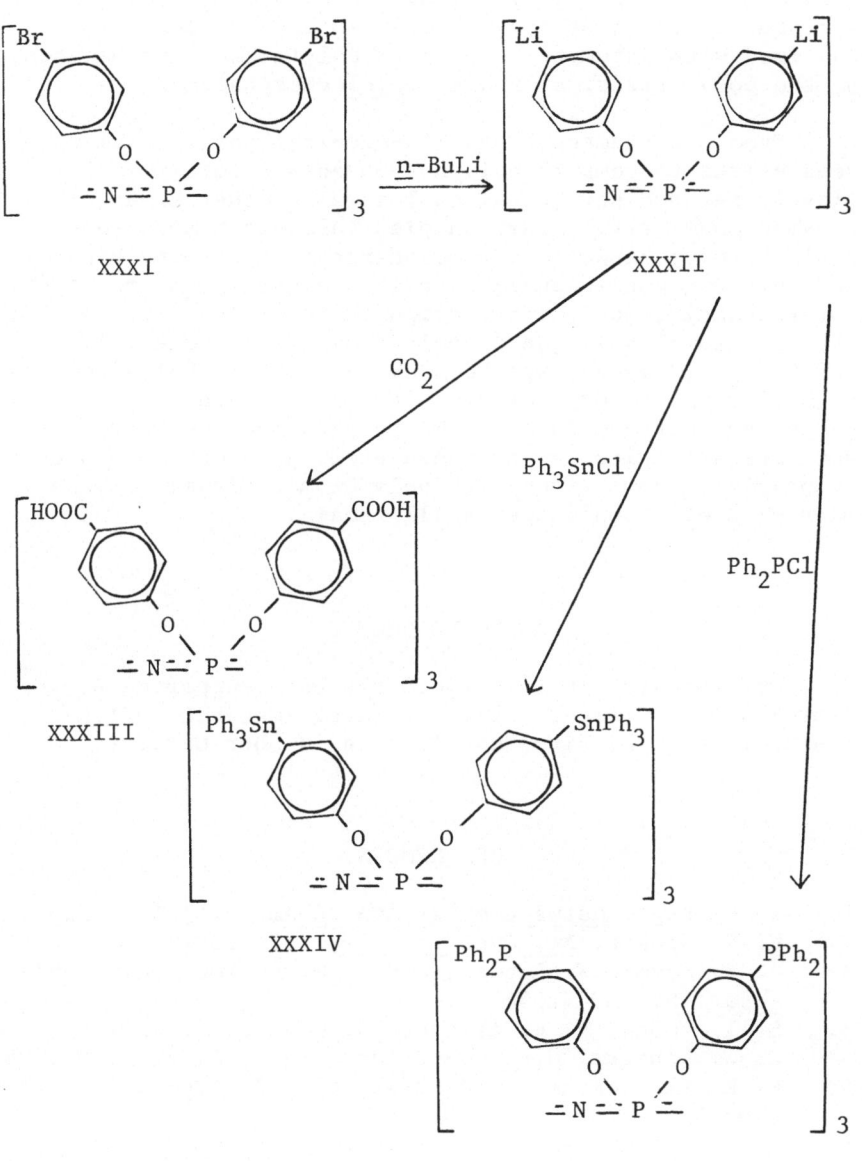

XXXI

XXXII

XXXIII

XXXIV

XXXV

metallic reagents or biologically-active compounds with polyphosphazenes.   Entirely new methods of polymer synthesis need to be developed.   We are currently involved with attempts to prepare poly(organophosphazenes) by the polymerization of organocyclophosphazenes.

From a fundamental physico-chemical point of view, a need exists to compare polyphosphazenes directly with closely related structural analogues in other areas of polymer chemistry.   For example, valuable comparisons could be made between polyphosphazenes on the one hand and poly(organosiloxanes) or poly(aldehydes) on the other. Unfortunately, such direct comparisons are not yet possible. The structural analogue of poly(dimethylsiloxane), i.e. $[NP(CH_3)_2]_n$, has not yet been synthesized.   Polysiloxanes with alkoxy, aryloxy, or amino residues as the sole side groups are not available.   No polyphosphazenes have yet been prepared with two hydrogen atoms on each phosphorus to provide a comparison with poly(oxymethylene).   Thus, many synthetic challenges still remain.

ACKNOWLEDGMENTS

Our research in this field has been supported by the Army Research Office, Office of Naval Research, N.I.H., N.A.S.A., and the Firestone Tire and Rubber Company.

REFERENCES

1.   J. Liebig, Justus Liebigs Ann. Chem., 11, 139 (1834).
2.   H. N. Stokes, Am. Chem. J., 19, 782 (1897).
3.   K. H. Meyer, W. Lotmar, and G. W. Pankow, Helv. Chim. Acta, 19, 930 (1936).
4.   E. T. McBee, H. R. Allcock, R. Caputo, A. Kalmus, and C. W. Roberts, U.S. Govt. Astia Rep., AD 209,669 (1959).
5.   H. R. Allcock and R. J. Best, Can. J. Chem., 42, 447 (1964).
6.   H. R. Allcock and R. L. Kugel, J. Am. Chem. Soc., 87, (1965).
7.   H. R. Allcock, R. L. Kugel, and K. J. Valan, Inorg. Chem., 5, 1709 (1966).

8. H. R. Allcock and R. L. Kugel, Inorg. Chem., 5, 1716
    (1966).
9. H. R. Allcock, J. E. Gardner, and K. M. Smeltz,
    Macromolecules, 8, 36 (1975).
10. H. R. Allcock, Phosphorus-Nitrogen Compounds, Academic
    Press, New York, 1972.
11. H. R. Allcock and C. Ting-Wah Chu, unpublished work
    (1978).
12. H. R. Allcock and R. J. Ritchie, unpublished work
    (1978).
13. H. R. Allcock, D. B. Patterson, and T. L. Evans,
    J. Am. Chem. Soc., 99, 6095 (1977).
14. H. R. Allcock, D. B. Patterson, and T. L. Evans,
    Macromolecules, submitted.
15. H. R. Allcock, W. J. Cook, and D. P. Mack, Inorg.
    Chem., 11, 2584 (1972).
16. H. R. Allcock, T. J. Fuller, D. P. Mack, K. Matsumura,
    and K. M. Smeltz, Macromolecules, 10, 824 (1977).
17. H. R. Allcock, S. D. Wright, and K. M. Kosydar,
    Macromolecules, 11, 357 (1978).
18. S. H. Rose, J. Polymer Sci., Part B, 6, 837 (1968).
19. D. P. Tate, J. Polymer Sci., Polym. Symp., 48, 33
    (1974).
20. G. S. Kyker, T. A. Antkowiak, Rubber Chem. Technol.,
    47, 32 (1974).
21. G. Allen, C. J. Lewis, and S. M. Todd, Polymer, 11
    31, 44 (1970).
22. R. E. Singler, N. S. Schneider, and G. L. Hagnauer,
    Polymer Eng. Sci., 15, 321 (1975).
23. R. E. Singler, G. L. Hagnauer, N. S. Schneider,
    B. R. LaLiberte, R. E. Sacher, and R. W. Matton,
    J. Polymer Sci., (Chem.), 12, 433 (1974).
24. G. L. Hagnauer and N. S. Schneider, J. Polymer Sci.,
    A-2, 10, 699 (1972).
25. J. E. White, R. E. Singler, and S. A. Leone, J.
    Polymer Sci., (Chem.), 13, 2531 (1975).
26. J. E. White and R. E. Singler, J. Polymer Sci.,
    (Chem.), 15, 1169 (1977).
27. N. S. Schneider, C. R. Desper, and J. J. Beres, in
    Liquid Crystalline Order in Polymers, (A. Blumstein,
    Ed.), Academic Press, New York, 1978.
28. H. R. Allcock, T. J. Fuller, and K. Matsumura,
    unpublished work (1978).
29. H. R. Allcock and G. Y. Moore, Macromolecules, 5,
    231 (1972).

30. H. R. Allcock, R. W. Allen, and J. P. O'Brien, J. Am. Chem. Soc., 99, 3984 (1977).

31. H. R. Allcock, P. P. Greigger, J. E. Gardner, and J. L. Schmutz, J. Am. Chem. Soc., in press.

32. The Firestone Tire and Rubber Company, Akron, Ohio.

33. J. E. Thomson and K. A. Reynard, J. Appl. Polymer Sci., 21, 2575 (1977).

34. J. P. O'Brien, W. T. Ferrar, and H. R. Allcock, Macromolecules, in press.

35. C. W. R. Wade, S. Gourlay, R. Rice, A. Hegyeli, R. Singler, and J. White, in Organometallic Polymers, (C. E. Carraher, J. E. Sheats, and C. U. Pittman, Eds.), Academic Press, New York, 1978.

36. D. P. Craig, M. L. Heffernan, R. Mason, and N. L. Paddock, J. Chem. Soc. (London), 1376 (1961).

37. M. J. S. Dewar, E. A. C. Lucken, and M. A. Whitehead, J. Chem. Soc. (London), 2423 (1960).

38. H. R. Allcock, R. L. Kugel, and E. G. Stroh, Inorg. Chem., 11, 1120 (1972).

39. H. R. Allcock, R. A. Arcus, and E. G. Stroh, unpublished work.

40. H. R. Allcock, R. W. Allen, and J. J. Meister, Macromolecules, 9, 950 (1976).

41. R. W. Allen and H. R. Allcock, Macromolecules, 9, 956 (1976).

42. T. L. Evans, T. J. Fuller, and H. R. Allcock, J. Am. Chem. Soc., in press.

LIQUID CRYSTALLINE ORDER IN POLYMERS CONTAINING ELEMENTS OF MESOGENIC STRUCTURE

Alexandre Blumstein

Department of Chemistry, Polymer Program
University of Lowell
Lowell, MA   01854

INTRODUCTION

Polymers which spontaneously, in bulk or solution, generate liquid crystalline domains have a positional and orientational molecular order which is intermediate between that of molecular crystals and of liquids.

Among low molecular weight compounds two large groups of mesophases can be distinguished, both generated by a weakening of intermolecular forces. In one group it is thermal energy which provides the desired effect (thermotropic mesomorphism) in the other, liquid crystalline organization arises through solvation (lyotropic mesomorphism). Both groups of mesophases can be observed in polymeric systems and both have received considerable attention in recent years. An industrially important example is given by high strength fibers spun from lyotropic solutions and melts of rigid chain polyamids and polyesters (1). Other important materials are micro-phase composites generated by block and graft diphilic copolymers (2, 3). Such systems have been abundantly reviewed and will not be discussed here. Liquid crystalline order in polypeptides (4) and in biological materials (5) has also been the subject of recent review articles and will not be within the scope of this article.

This paper is primarily concerned with polymers displaying thermotropic mesomorphism. The
moieties which generate the liquid crystalline
organization are called the mesogenic moieties
and can be found either in the main chain or in
the side groups. The structural characteristics
associated with liquid crystalline behavior have
been reviewed by Gray (6, 7) and will not be discussed here.

Table I summarizes some general features of
the mesomorphic order in liquid crystalline mesophases. In polymers as well as low molecular
materials the mesomorphic order is characterized
by the nature of the positional and orientational
order of segments (or molecules) within an oriented domain. The supermolecular structure of the
mesomorphic materials is characterized by the positional (and sometimes orientational) order of
the domains themselves. In the case of liquid
crystals, in contrast to crystals, the molecular
(or segmental) positional order is lost to some
degree while the orientational order subsists.
The extent of loss of positional order determines
the nature of the liquid crystal.

The relationship between polymers and liquid
crystals began with a series of investigations in
which liquid-crystalline arrangements of monomer
molecules were used as polymerization media (8-15).
These studies were primarily concerned with the
topotactic and topochemical effects which one
might expect to be associated with polymerization
within the oriented liquid crystalline matrix.
Among the first synthetic systems studied were
tightly crosslinked glasses with "locked in" liquid crystalline order (17). Typical monomers
used to prepare these glasses have two vinyl
groups located at both ends of a rigid mesogenic
moiety such as for example Di(n-p.acryloyloxyben-
zylidene)-p.diaminobenzene:

TABLE 1. General Features of Order in some Thermotropic Mesophases

Mesophase	Order of molecules (segments)		X-ray Features	
	Orientational	Positional	Unoriented Samples	Oriented Samples
Nematic	Orientation around the director: $Q=(3\cos^2\theta-1)/2$	Random	Weak diffuse inner ring. Strong diffuse outer ring.	Splitting of both rings.
Smectic A	Parallel to layer normal	Random within layers	One (or more) sharp inner reflections. One diffuse outer ring.	Splitting of inner ring around the meridien. Equatorial outer ring splitting. Optically uniaxial.
Smectic C	Parallel to each other. Tilted with respect to layer normal	Random within layers	Similar to Sm A	Splitting of inner ring around meridien. Outer ring splitting into equatorial doublet.
Smectic B	Parallel to layer normal	Two dimensional hexagonal	Sharp inner and outer rings.	Splitting of both rings similar to Sm A
Smectic B$_t$	Mutually parallel, tilted with respect to layer normal	Two dimensional hexagonal	Similar to Sm B	Splitting of both rings similar to Sm C

$$CH_2=CHCOO-\langle O \rangle-CH=N-\langle O \rangle-N=CH-\langle O \rangle-OOCCH=CH_2$$

Remarkably, Bouligand, Liebert, Strzelecki and co-workers found that in such systems the mesomorphic texture of the monomer can be locked into the polymeric glass down to the slightest disclination (16, 17).

More recent work has shown (18, 19, 20) that glasses with "locked in" liquid crystalline order can be obtained even in the absence of a crosslinking divinyl compound. This order is not necessarily related to the mesomorphic order of the monomer and, furthermore, the monomer itself need not be mesomorphic (20, 21) although a latent mesogenic structure characterized by rigidity and anizotropy of various properties is necessary. Furthermore, the polymer need not be prepared within a liquid crystalline matrix. The driving force in the formation of liquid crystalline order is provided by the mesogenic group itself, whether incorporated in the side group or present in the main chain. Polymeric mesophases can then be obtained by providing the polymer with the necessary segmental mobility (20). The majority of such polymers display high glass transition temperatures and their liquid crystal/isotropic transition temperatures and enthalpies are difficult to observe because they often overlap with the region of chemical instability of the polymer. Recently a range of polymers have been prepared in which the mesogenic group was separated from the main chain by means of a sequence of 6-12 methylene units (flexible spacers) (22, 23, 24, 25). Such polymers undergo reversible, thermotropic phase transitions, comparable to that of low molecular weight mesogenes.

Since 1973 the number of papers dealing with mesogenic monomers and polymers has multiplied. Synthetic methods and some structure property correlations have been reviewed in a recent book (26). A structural classification of liquid crys-

talline order in polymers based on X-ray diffraction has also been presented (27).

A number of questions, however, remain large-
ly unresolved even from a qualitative point of
view. The original question concerning possible
topotactic and topochemical effects favored by the
use of liquid crystals as polymerization media,
has yet to be fully answered. Much less have been
answered questions concerning the role of the
structure of the mesogenic groups, the role and
flexibility of the backbone, the role of molecular weight in order formation. The mechanisms of
nucleation, growth and alignment of mesomorphic
domains in the polymer remain to be studied inspite of the importance of these questions in
practical applications.

In this paper some of these questions will
be discussed in the light of currently available
experimental evidence.

DISCUSSION

The Mesogenic Moiety and
the Liquid Crystalline Order

The structure of the mesogenic moiety appears
to exert a strong influence on the nature of liquid crystalline order developed in a polymer.
In low molecular weight compounds the views concerning the rules of molecular mesomorphism are
being refined as more data are becoming available.
It is argued today that geometric factors play
the determining role in the nature of interaction
between mesogenic molecules (28). It is agreed
that the forces maintaining order in the nematic
state are primarily dispersion forces while forces
maintaining molecular cohesion in the smectic
state can be, in addition, polar in nature. Thus,
smectic order is more dependent on dipole-dipole
interactions than the nematic order. In this
context it appears clear that dipole moments

operating across the molecular axes are promoting
lateral cohesion and, if a suitable geometric
packing can be achieved, the smectic order is enhanced. Conversely, strong dipole moments along
molecular axes have a destabilizing effect on
smectic mesophases through simple electrostatic
repulsive action.

In the case of vinyl polymers the mesogenic
side groups are confined to each other's vicinity
and the natural predisposition to clustering between the strongly interacting mesogenic moieties
is therefore enhanced. If the geometry requirements are satisfied to provide a close packing of
such moieties, a smectic, layered organization
results. A flexible backbone facilitates formation of such arrays. One distinguishes several
organizations leading to layered structure in
polymers:

  (1) Straight double layers corresponding
       to Smectic A or B arrangements (29, 22).
  (2) Interpenetrating double layers (22).
  (3) Tilted arrangements (20, 30).

Most polymers with mesogenic moieties built
into the side group are dominated by lamellar
arrangements both below (61) and above (22,
24) the glass transition temperature. A certain
number of nematic polymeric structures have nevertheless been reported. Such structures are
the result of destabilization of the lateral coupling between mesogenic units either because of
dipole repulsion, dipole weakening or unfavorable
geometry which does not allow lateral dipole moments to interact efficiently.

Polymers with rigid mesogenic segments in
the backbone display in the lyotopic state a nematic mesophase in agreement with the earlier
theoretical predictions (31, 32). In the melt,
however, such structures have not been investigated extensively because of forbiddingly high flow
temperatures. Linear copolymers containing variable contents of poly(ethyleneterephtalate) and

rigid segments containing oxybenzoyl moieties have been successfully studied by the Eastman Kodak Group (33, 34). It was found that melts of copolyesters with high content of oxybenzoyl units displayed spontaneous nematic ordering and a remarkable enhancement of mechanical properties in the direction of the chain ordering.

For polymers in which the incorporation of flexible sequences between the rigid mesogenic moieties allows for lower transition temperatures some recent X-ray data also indicate a prominence of smectic, layered structures (23, 25). In addition to X-ray patterns, electron diffraction performed on films of the copolymer of low molecular weight of p-hydroxybenzylidene-p.aminophenol and sebacic acid indicates the existence of a lamellar lattice with hexagonal symmetry suggesting the possibility of a Smectic B_t (tilted) disposition of chains (25). This tendency of linear polymers with rigid mesogenic segments separated by a sequence of methylene units to form smectic domains brings up the question of the role of the backbone and of its flexibility. At present there is not enough crystallographic data on such systems to explain the formation of lamellar structures. One can only speculate that this may be due to the introduction of transverse carboxylic dipoles or possibly to a local "microphase separation" between the flexible methylene spacers and the rigid polar mesogenic groups.

The Role of Tacticity

When in 1967 it was stated by Amerik and Krentsel (35) that vinyloleate can be polymerized in bulk in its smectic state to produce a stereoregular polymer, it was an exciting suggestion along the lines of Kargin's concept of polymerization through "oriented monomer sequences." The evidence for this statement was based on the relatively sharp X-ray pattern obtained from poly(vinyloleate) by polymerizing the monomer in its smectic phase. Subsequent studies of free

radical polymerization of and within mesophases
have shown that there is no significant change in
the tacticity of polymers obtained (20, 30, 36,
37).   Organized smectic domains can form spontaneously from atactic polymers.   This has been
abundantly demonstrated with different monomers.
Table 2 lists some polymers which have been shown to
form layered, smectic domains, the inter-lamellar
spacing being independent of the tacticity.   It
appears that the tendency of atactic polymers with
interacting side groups to organize reaches beyond
the realm of smectic ordering and into three di-
mensional positional and orientational order.
Polymers such as poly(acroloyloxybenzoic acid)
(30), poly(vinyltrifluoroacetophenone) (38), and
poly(vinylcarbazole) (39) have been recently
shown to crystallize inspite of a proven atactic
configuration.

        In all cases mentioned above, the side group
interactions are strong enough to distort the
backbone sufficiently to favor side group stack-
ing. Not surprisingly, backbone flexibility appears to be an important parameter on which the
extent of segmental positional and orientational
order depends.   The example of poly(methacryloyloxybenzoic acid) which has a very low tendency
to crystallize (but easily gives smectic, layered
domains) and of poly(acryloyloxybenzoic acid)
which crystallizes easily is a case in point.

        These findings require us to reexamine the
concepts held for decades and according to which
stereoregularity (in a chain with protruding side
groups) is the necessary condition for polymer
crystallization.   A detailed crystallographic
analysis of such polymers is a prerequisite in
understanding their crystallization mechanism but
is at present lacking.

TABLE 2.   Examples of Liquid Crystalline Order
           in Atactic Polymers

Polymer	Fraction of Isotactic dyads	Spacings $\overset{o}{dA}$ $\overset{o}{DA}$		Reference
$\{CH-CH_2\}_x$ $\mid$ $O=C-O-$(biphenyl)	.89 .62	23.2 23.2	5.2 5.2	(36) (36)
CH_3 $\{C-CH_2\}_x$ $\mid$ $O=C-O-$(phenyl)$-COOH$	.28	18.5	4.4	(20)
CH_3 $\{C-CH_2\}_x$ $\mid$ $O=C-O-Chol.$	.25 .75	35.3 35.3	6.35 6.35	(21) (21)
CH_3 $\{C-CH_2\}_x$ $\mid$ $O=C-O-$(phenyl)$-CH=N-$(phenyl)$-COOH$	.25	32	5.1	(50)

Liquid Crystalline Order
and Molecular Weight

Molecular weight can affect the liquid-crystalline order of polymeric melts. It is known
that molecular weight can affect the intra-molecular order in a worm-like chain. For a worm-like
chain Tsvetkov, et al. (40) established the relation between Q, the degree of orientational axial
order and x, the ratio of contour length L to
persistence length a:

$$Q = \frac{6/5 \ [1-(1-e^{-x})\frac{1}{x}]}{x-0.8 \ [1-(1-e^{-x})\frac{1}{x}]}$$

In this expression Q decreases monotonically as x
increases and for a given backbone rigidity Q decreases with increasing length of the macromolecule. This equation was shown to be a good approximation for a number of rigid and semi rigid
chain polymers in dilute solution. The intermolecular order which is established through an extensive network of contacts between individual
macromolecules will thus be also affected as well
as the size of the ordered domains. In the case
of more flexible vinyl polymers with mesogenic
side groups the main chain plays a more subdued
role. It is not expected to affect significantly
the intramolecular order. The intermolecular
order can be affected for very low molecular
weights due to fragmentation of domains, but a
systematic investigation of the influence of the
molecular weight on order is yet to be done.

In comblike molecules with long hydrocarbon
backbones, especially grafted homopolymers, the
intramolecular axial order increases with M as
reflected by studies of flow birefringence of
such molecules (40). In this case the intramolecular order increases sharply with the length of

the branches and with their number per unit backbone length.

Compatibility Between Polymers
with Mesogenic Segments
and Low Molecular Weight
Liquid Crystals

At present there is little data on identification of polymeric mesophases by microscopy.
X-ray, electron scattering and thermal analysis
remain the main tools in analyzing liquid crystalline order in condensed polymeric materials. The
effectiveness of optical microscopy and of the
compatibility rules formulated by Sackmann and
Demus a decade ago and widely used in identification of mesophases formed by low molecular compounds (41, 42) is questionable in the case of
high molecular weight materials. The compatibility rules state that mesophases of similar nature
should display total miscibility. However, polymers are particularly difficult to mix with liquid
crystalline solvents. This is due to the very low
gain in entropy in such solutions and to the diphilic nature of the polymer (a polar side group attached to a hydrocarbon backbone). In addition,
the high viscosity of mesomorphic solvents and of
the polymeric chain makes the establishment of
thermodynamic equilibrium difficult and increases
the difficulty of obtaining the characteristic
textures widely used in the identification of low
molecular weight mesophases. Nevertheless, Cser
et al. (43) claim total miscibility of p-alkyl-p'-
acryloyloxyazoxybenzenes

$$CH_2=CH-COO-\langle\bigcirc\rangle-N\underset{O}{\overset{\uparrow}{=}}N-\langle\bigcirc\rangle-R$$

with the corresponding polymers in the nematic
state and bring, therefore, an argument in favor
of compatibility rule for polymeric systems. However, the lack of compatibility between a polymer
such as poly(methacryloyloxybenzoic acid) charac-

terized by a smetic C organization (20, 27) and
the corresponding alkoxybenzoic acids in their
smectic C state (54) is difficult to explain in
terms of this rule.

Recently, Dubault et al. (44) tried to dissolve low molecular poly(ethyleneglycol) in the
nematic phase of p.-azoxyanizole. Only very low
molecular weights (2,000) could be dissolved up
to a maximum of 1% (wt.) concentration. In contrast Noel and Billard (45) have reported complete miscibility of a poly(azomethine) of a low
molecular weight (2,000-4,000) with three nematic
reference solvents. The nematic solvents and the
nematic polymer were characterized by a very similar molecular structure translated into similar
solubility parameters, making it the first successful example of application of the compatibility
rules to polymeric systems. Substantially more
data is needed to answer the question of effectiveness of compatibility rules in the field of
macromolecules.

Domain Formation and Alignment

The intrinsic tendency to organize is well
pronounced in most polymers with mesogenic side-
groups. This realization has come about very recently. Following the earlier work of Liebert
and Strzelecki it was thought that the presence
of a cross-linking agent is necessary to "lock
in" the mesomorphic structure and texture of the
monomer. (See Ref. 17 and related papers). The
authors had carried out their experiments at very
high polymerization rates under conditions such
that the time of segmental relaxation was much
longer than the time between successive additions
of monomer units and an effective locking through
crosslinks was assured. The order which developed was usually, though not always, the very
same order as that of the monomer (29).

It was shown later (18, 19, 20, 21) that
mono-functional monomers can also spontaneously

give polymers with liquid crystalline order. The
necessary condition appears to be the existence
of elements of mesogenic structure in the monomer.
These elements interact to nucleate domains in
which liquid crystalline order prevails. This
order is not necessarily related to that of the
monomer. Polymers with mesogenic side groups
sometimes show a tendency to give random conformations in the nascent state and on precipitation
from good solvents. Annealing, however, enhances
the ordering. In such polymers the backbone promotes the formation of layered arrays but is a
disturbing element in the formation of the nematic order.

The development of ordered domains in polymers in bulk is strongly dependent on the compatibility of the polymer and monomer and/or solvent. It has been shown by Tsvetkov et al. (40)
that the nature of the solvent affects the intramolecular order and hence the conformational and
morphological properties of the macromolecule.

Table 3 shows the dependence of segmental
optical anizotropy $\Delta\alpha$ on the solvent quality
(as given by the value of the limiting viscosity
number $[\eta]$) for the polymer of the p-phenylme-
thacryloylester of p-cetyloxybenzoic acid. Since
$\Delta\alpha$ is directly related to the degree of intramolecular orientational order, one can see that
the latter decreases in good solvents (high $[\eta]$).

External fields of force such as mechanical,
hydrodynamic, magnetic and surface forces can enhance order and increase size and orientation of
the domains. Fields of force can also induce a
nematic ordering in non-mesogenic polymers, but
this nematic orientation is metastable and disappears as soon as the orienting field is removed
and the polymer allowed to relax (heating, annealing).

Polymers in which the mesogenic group is
connected to the main chain via a flexible spacer
display liquid crystalline textures characteristic

TABLE 3. Dependence of Segmental Optical Anizotropy $\Delta \alpha$ on the Solvent Quality for the Polymer of the p.phenylmethacryloylester of p.cetyloxybenzoic acid

SOLVENT	$[\eta]$ dl/g	$\Delta \alpha \ 10^{25} (cm^3)$
Tetrahydrofuran	3.08	-890
Chloroform	2.88	-1400
Benzene	2.50	-1600
Tetrachloromethane	0.63	-2700
Benzene and Heptane	0.54	-4200

*Data from reference #40

of large submicroscopic domains such as planar
textures (22) and Grandjean textures (46). This
may be a consequence of "decoupling" the mesogenic group from the backbone with a concomitant
drop in local viscosity and increase of segmental
mobility.

The size of the mesomorphic domain can vary
considerably, from about 100Å in unoriented polymer to microscopic dimensions in strongly oriented
samples (29, 47, 51). Recent analysis of the
shape of the X-ray low angle scattering intensity
distribution (48) performed on unoriented poly
(cholesterylmethacrylate) seems to indicate domain
sizes varying between 100Å - 1000Å depending on
the sample preparation method.

Alignment of domains in mesogenic polymers
can be achieved either by polymerizing an aligned
monomer or by submitting the mesogenic polymer to
a directional deformation in a field of force.
The first method was used with divinyl monomers
at high temperatures by locking in the orientation of the monomer (29, 49) produced by a magnetic field of 8000 - 10,000 Oe. It was also successfully used with monovinyl monomers either
oriented by a magnetic field (47, 27) or oriented
by surfaces (50, 51). However, in a number of
experiments alignment of mesomorphic domains was
not obtained inspite of the alignment of the mesomorphic monomer molecules. The factors on which
the successful outcome of such alignment experiments depend are not well understood.

The second method involving orientation of
mesogenic thermotropic polymers in external force
fields was not described extensively in the literature. The advent of mesogenic, thermotropic
polymers with flexible spacers (22, 23, 24, 25)
displaying relatively low transition temperatures
will doubtlessly facilitate experimentation in
this area.

Liquid Crystalline Phases as Media
for Free Radical Polymerization

Experimental work in kinetics of polymerization in liquid crystalline media is sketchy at best. Hopes have been formulated for the possibility of regulating stereo-placements and inducement of topotactic effects by free-radical polymerization of liquid crystalline monomers in bulk or in liquid crystalline solvents, due to the high degree of orientational order (35, 52). Such effects have yet to be established. Most of the reported data appear to be due to factors other than molecular orientation and are only indirectly related to the liquid crystalline order of the medium. For example, factors such as incomplete miscibility of monomer and solvent, phase separation of the polymer, enhancement of viscosity of the medium, and caging of molecules of initiator (53, 54, 55, 56) can be invoked to explain the observed kinetic effects.

A detailed kinetic analysis of monomers with strongly interacting side groups (57) has shown that the rate enhancement observed with such monomers is due to a decrease in the termination rate K_t. One can see from Table 4 that the higher the anizotropy of polarizabilities $\Delta\alpha$ of the statistical segment in the polymer chain, the lower K_t. Since, on the other hand, $\Delta\alpha$ is related to the order parameter Q, the authors explain the observed decrease in K_t as due to the development of intramolecular order between the mesogenic side groups along the backbone and the corresponding increase in stiffness of the backbone translating into a reduced mobility of the radical chain ends.

There is little experimental evidence to support various speculative claims that liquid-crystalline order leads to enhancement of the propagation rate constant, such as proposed earlier by Kargin and his school in his model of "oriented aggregates" in which polymerization rate could proceed with an enhanced collision factor (58).

TABLE 4.   Absolute Rate Constants and Segmental Anizotropy for Some Methacrylic Esters

MONOMER	$t\,^{\circ}C$	$K_p^{(*)}$	$K_t^{(*)} \cdot 10^{-7}$	$\Delta_\alpha^{(**)} \cdot 10^{25}\,cm^3$
Methylmethacrylate	50	580	6.9	2
n–butylmethacrylate	30	362	1.0	-14
Cetylmethacrylate	50	–	–	-160
Methacryloyloxyphenylester of cetyloxybenzoic acid in				
toluene	50	170	0.03	–
dioxane	50	300	0.025	-3100

(*) Data from reference #57

(**) Data from reference #40

It must, however, be pointed out that most of the
kinetic work was done in the simple nematic phase
in which the degree of positional molecular order
is very low (59, 60). Little systematic work has
been done in smectic phases, in which frequent
phase separation of the forming polymer and high
viscosities are a complicating factor.

Copolymers

Copolymers of mesogenic with non-mesogenic
vinyl monomers provide the opportunity to diver-
sify the properties of polymers with liquid crystalline order.

As in the case of homopolymers, the determination of optical segmental anizotropy $\Delta \alpha$ provides a tool for the investigation of the intramolecular order in the copolymer in solution. The
dependence between $\Delta \alpha$ and the copolymer composition was predicted by Tsvetkov et al.(40) who have
shown it to be non linear. The steepness of the
decline of $\Delta \alpha$ with decreasing composition of the
mesogenic comonomer was shown to depend on flexibility and on the difference in optical anizotropy
between the two comonomer units. These considerations were quantitatively verified on the example
of a statistical copolymer the methacryloyloxy-
phenylester of the cetyloxybenzoic acid with cetyl-
methacrylate (40).

In our work with copolymers of cholesteryl
methacrylate with a number alkylmethacrylates we
have shown (21) that these considerations are
qualitatively valid also for copolymers in their
condensed state. It was found that for a random
copolymer of a fixed composition the degree of
organization dramatically decreased when side
groups of the non-mesogenic comonomer unit inter-
fered through steric hindrance with the formation
of ordered arrangements of cholesteric side groups.

This effect was also shown by Shibaev et al.
(22) who studied copolymers of various alkylme-

thacrylates with cholesterolesters of N-methacry-
loyl-ω-aminocarboxylic acid: $CH_2=C(CH_3)-CONH-(CH_2)_n-$
COOChol with different lengths of the aliphatic
radical between the amino and carboxy groups
(n=2 to 11). The high flexibility of the methylene bridge connecting the cholesterol moiety to
the backbone lowers the order-disorder transition
temperature of the side group and gives a direct
opportunity to observe reversible enantiotropic
phase transitions. It was also shown by the same
authors that for a bridge of eleven methylene
units, the copolymer with alkylmethacrylates can
still preserve its liquid crystalline order in
bulk down to 40 mole % of the cholesteric monomer
if the comonomer is n-butylmethacrylate. If the
length of the alkyl group in the methacrylic unit
is increased, the stability of the mesophase tends
to decrease. A copolymer in which the alkylradi-
cal has 22 methylene units does not present any
anizotropy whatever the composition.

Even more intriguing are the results of
Finkelmann et al.'(24, 46) who by proper section
of comonomers containing either a cholesterol
moiety or an assymetric carbon atom (methylbenzy-
lamino group) incorporated into a mesogenic moiety succeeded in synthesizing an uncrosslinked
cholesteric polymer. The first system (24) was
obtained by co-polymerization of two monomers in
which the cholesteric group was placed at the end
of a methylene bridge

$$CH_2=C(CH_3)-COO-\left(\bigcirc\right)-(CH_2)_n-COOChol$$

The methylene bridge was short (n=2) for one of
the comonomers and long (n=12) for the other.
The mismatch in the positioning of mesogenic moieties along the chain disturbed the layered arrangement of side groups inducing a (twisted) nematic
order of the mesogene. This spontaneous formation
of twisted nematic phases in the copolymer would,
however, be obtained only within a sharp interval
of compositions (.5 mole fraction) outside of

which a layered (smectic) arrangement of side
groups was observed. In the second system two
monomers were used both giving smectic and nematic mesophases:

$$CH_2=C(CH_3)COO(CH_2)_6O\bigcirc COO\bigcirc\bigcirc OCH_3$$

$$CH_2=C(CH_3)COO(CH_2)_2O\bigcirc COO\bigcirc CH=N^*CH(CH_3)Ph$$

The copolymerization gave copolymers displaying
twisted nematic structures in a large interval of
compositions. Grandjean textures and accompanying
selective reflections of colors are easily obtained with these interesting new materials.

It is apparent from this overview that the
field of liquid crystalline polymers is still in
its early stages of development but is rapidly
growing. The study of liquid crystalline polymers
is central to materials' science: it involves the
study of crystalline polymers in the regions of
intermediate order and the nature of nucleation
and crystal growth in polymeric melts and solutions which often form from initially liquid crystalline phases. Just as interesting is the study
of intra-molecular order in mesogenic macromolecules and their inter-molecular association in
concentrated solutions and melts. Such studies
go beyond the development of ultra-high modulus
fibers in promoting a better understanding of the
formation of liquid crystalline morphoses of living tissues and enhance the prospect of generating
synthetic polymeric materials closely ressembling
the natural materials of the biosphere.

ACKNOWLEDGEMENT

Partial support of the National Science
Foundation under the Research Grant DMR 75-17397
is gratefully acknowledged. Acknowledgement is
made to the donors of Petroleum Research Fund

administered by the American Chemical Society for partial support under the PRF Research Grant 8285AC6C. Thanks are expressed to Dr. Rita B. Blumstein for stimulating discussions.

REFERENCES

1. J. Preston in "Liquid Crystalline Order in Polymers," A. Blumstein Editor, Academic Press Pbsh., 1978.
2. A. Skoulios in Advances in Liquid Crystals $\underline{1}$, 169 (1975) G.H. Brown, Editor, Academic Press Pbsh.
3. B. Gallot in "Liquid Crystalline Order in Polymers," A. Blumstein, Ed., Academic Press Pbsh., 1978.
4. E. Samulski in "Liquid Crystalline Order in Polymers," A. Blumstein Editor, Academic Press Pbsh., 1978.
5. Y. Bouligand in "Liquid Crystalline Order in Polymers," A. Blumstein, Editor, Academic Press Pbsh., 1978.
6. G.W. Gray, Molecular Structure and the Properties of Liquid Crystals, Academic Press, 1962.
7. G.W. Gray, "Molecular Geometry and the Properties of Nonamphiphilic Liquid Crystals," in "Advances in Liquid Crystals" $\underline{2}$, p.1, 1976, G.H. Brown Editor, Academic Press Pbsh.
8. A. Blumstein, Bull. Soc. Chem. Fr. 899, 906 (1961).
9. J. Herz, F. Reiss-Husson, P. Rempp and V. Luzzati, J. Polymer Sci.C,1275 (1963).
10. Ch. Sadron, Angew. Chem. Int. Ed. 75, 11, 472 (1963).
11. A. Blumstein, S.L. Malhotra and A.C. Watterson, J. Polymer Sci. A-2, $\underline{8}$, 1599 (1970).
12. A. Blumstein, R. Blumstein, T.H. Vanderspurt, J. Colloid and Interface Sci., $\underline{31}$, 236 (1969).
13. Y.B. Amerik and B.A. Krentsel, J. Polymer Sci. C16, 1383 (1967).
14. Y.B. Amerik, I.I. Konstantinov and B.A. Krentsel, J. Polymer Sci. $\underline{C23}$, 231 (1968).

15.   H.Z. Friedlander and C.R. Frink, J. Polymer
      Sci. B2, 475 (1964).
16.   Y. Bouligand, P.E. Cladis, L. Liebert and
      L. Strzelecki, Molecular Cryst. Liquid
      Cryst. 25, 233 (1974).
17.   L. Liebert and L. Strzelecki, Bull. Soc.
      Chim. Fr., p. 603 (1973).
18.   P.L. Magagnini, A. Marchetti, F. Matera,
      G. Pizzirani and G. Turchi, European Polymer
      J., 10, 585 (1974).
19.   E. Perplies, H. Ringsdorf and J.H. Wendorff,
      Makromol. Chem. 175, 553 (1974).
20.   A. Blumstein, R. Blumstein. S.B. Clough and
      E.C. Hsu, Macromolecules, 73 (1975).
21.   A. Blumstein, Y. Osada, S.B. Clough, E.C. Hsu
      and R.B. Blumstein in "Mesomorphic Order in
      Polymers and Polymerization in Liquid Crys-
      talline Media" A. Blumstein, Editor, ACS Sym-
      posium Series 74, 56 (1978).
22.   V.P. Shibaev, N.A. Plate and Y.S. Freidzon,
      in "Mesomorphic Order in Polymers and Poly-
      merization in Liquid Crystalline Media" A.
      Blumstein, Editor, ACS Symposium Series, 74
      p.33 (1978).
23.   A. Roviello and A. Sirigu J. Polymer Sci.,
      (Polymer Lett.) 13, 455 (1975).
24.   H. Finkelmann, H. Ringsdorf, W. Siol and J.H.
      Wendorff in "Mesomorphic Order in Polymers
      and Polymerization in Liquid Crystalline
      Media," A. Blumstein, Editor, ACS Symposium
      Series 74, p.22 (1978).
25.   K.N. Sivaramakrishnan, A. Blumstein, S.B.
      Clough and R.B. Blumstein, ACS Polymer
      Preprints 19, 2, 190 (1978).
26.   A. Blumstein, Editor "Liquid Crystalline
      Order in Polymers", Academic Press Pbsh.,
      1978.
27.   S.B. Clough, A. Blumstein and A. deVries in
      "Mesomorphic Order in Polymers and Polymeri-
      zation in Liquid Crystalline Media," A.
      Blumstein Editor, ACS Symposia V. 74, p. 1
      (1978).
28.   J. Malthete, J. Billard, J. Canceill,
      J. Galard et J. Jacques, J. Physique C3, 37,
      p. 1 (1976).

29.   S.B. Clough, A. Blumstein and E.C. Hsu, Ma-
      cromolecules 9, 123 (1976).
30.   A. Blumstein, S.B. Clough, L. Patel, R.B.
      Blumstein and E.C. Hsu, Macromolecules 9,
      243 (1976).
31.   P.J. Flory, Proc. Roy. Soc. A234, 73 (1956).
32.   E.A. diMarzio, J. Chem. Phys. 35, 658 (1961).
33.   W.J. Jackson Jr. and H.F. Kuhfuss, J. Poly-
      mer Sci., Chem. Ed., 14, 2043 (1976).
34.   F.E. McFarlane, V.A. Nicely and T.G. Davis
      in Proceedings of the 8th Biennial Sympo-
      sium of Division of Polymer Chemistry, J.R.
      Schaefgen and E.M. Pearce, Ed. Plenum Press,
      1977.
35.   Y.B. Amerik and B.A. Krentsel, J. Polymer
      Sci., J. Polymer Sci. C, 16, 1383 (1967).
36.   A. Cecarelli, V. Frosini, P.L. Magagnini and
      B.A. Newman, J. Polymer Sci. (Polymer Lett.),
      13, 101 (1975).
37.   Y. Osada and A. Blumstein, J. Polymer Sci.,
      (Polymer Lett.) 15, 761 (1977).
38.   S.C. Chu and E.M. Pearce, J. Polymer Sci.
      (Polymer Letters), 14, 375 (1976).
39.   C.H. Griffith, J. Polymer Sci. (Polymer
      Lett.) 16, 271 (1978).
40.   V.N. Tsvetkov, E.I. Riumtsev and I.N. Shtenni-
      kova, in "Liquid Crystalline Order in Poly-
      mers" A. Blumstein Editor, Academic Press.
      p. 43 (1978).
41.   J. Billard, Bull. Soc. Fr. Miner. 95, 206
      (1972).
42.   H. Sackmann and D. Demus, Mol. Cryst. Liq.
      Cryst. 21, 239 (1973).
43.   F. Cser, K. Nyitrai, E. Seyfried and Gy.
      Hardy Erop. Polymer J. 13, 679 (1977).
44.   A. Dubault, C. Casagrande and M. Veyssie
      Molecular Cryst. Liquid Cryst. (Letters) in
      print.
45.   C. Noel and J. Billard, Abstracts of papers
      presented at the 7th International Confe-
      rence on Liquid Crystals, Bordeaux July 1978.
46.   H. Finkelmann, M. Portugal and H. Ringsdorf,
      A.C.S. Polymer Preprints; 19, 2, 183 (1978);
      Makromol Chem. Short Com. 179, 2541 (1978).

47. J.H. Wendorff, E. Perplies and H. Ringsdorf,
     Prog. Colloid Sci. 57, 272 (1975).
48. S.B. Clough and P. Boduch (to be published).
49. L. Liebert and L. Strzelecki, C. R. Acad.
     Sci. Ser. C276, 647.
50. A. Blumstein., L. Lim, unpublished results.
51. H.J. Lorkowski and F. Reuther, Plaste Und
     Kautschuk, 23, 2, 81 (1976).
52. B.A. Krentsel, in "Polymerization of Orga-
     nized Systems" H.G. Elias, Editor, Gordon
     and Breach Pbsh. 1977, p. 117.
53. A. Blumstein, J. Billard and R. Blumstein
     Molecular Cryst. Liquid Cryst. 25, 83 (1974).
54. A. Blumstein, R. Blumstein, G.J. Murphy,
     C. Wilson and J. Billard, "Liquid Crystals
     and Ordered Fluids," J.F. Johnson and R.S.
     Porter Editors, Vol. 2, p. 277 (1974) Plenum
     Press Pbsh.
55. G.J. Murphy, M. Sci. Thesis, University of
     Lowell, 1974.
56. E. Perplies, H. Ringsdorf and J.H. Wendorff
     in "Polymerization of Organized Systems"
     H.G. Elias, Editor, Gordon and Breach Pbsh.
     1977 p. 149.
57. A.A. Baturin, Y.B. Amerik, B.A. Krentsel,
     V.N. Tsvetkov, I.N. Shtennikova and E.I.
     Ryumtsev, Doklady, Akad. Nauk SSSR, 202, 3,
     586 (1972).
58. V.A. Kargin, V.A. Kabanov and I.M. Papissov,
     J. Polymer Sci., C4, 767 (1963).
59. C.M. Paleos and M.M. Labes, Molecular Cryst.
     Liquid Cryst., 11, 385 (1970).
60. E.C. Hsu and A. Blumstein, J. Polymer Sci.
     (Lett.) 15, 129 (1977).
61. A. Blumstein, Macromolecules 10, 872 (1977).

FORMATION OF POLYMERIC MATERIALS BY A NONPOLYMERIZATION

PROCESS: GLOW DISCHARGE POLYMERIZATION

H. Yasuda

Department of Chemical Engineering

University of Missouri, Rolla, MO 65401

Recognition of the formation of polymeric materials in a plasma state (1874) precedes the establishment of the concept of "polymer". In early observations, polymeric materials were recognized as undesirable by-products associated with electric discharge. In recent years, this phenomenon has been accepted as a potentially important process of coating or polymerization, and today the research that has evolved encompasses a host of applications and fundamental aspects. The process is generally referred to as "glow discharge polymerization" or "plasma polymerization".

The study of glow discharge polymerization is greatly hampered by the fact that those polymers having advantageous characteristics for potential application are highly crosslinked and branched and are thereby insoluble and infusible. In other words, the very features, which make them usable, actually hamper their characterization and hinder the study of their polymerizing mechanisms. Consequently, most studies concerning them are based on a preconceived concept that their formation is a kind of polymerization. This situation is somewhat analogous to the basic concept of "radiation polymerization" or "radiation induced polymerization", in which the initial step is the direct consequence of the radiation process, but these polymer forming processes are essentially identical to those initiated by conventional means.

Recent studies tend to indicate that the polymer formation process in the plasma state may not be adequately

represented by the conventional concept of "polymerization". Because of this, the following review of the process is based on a completely different viewpoint so that
some insight may be gained of the mechanism of polymer
formation.

NON-POLYMERIZATION PROCESS

The formation of a polymer in glow discharge is a
complex phenomenon, but interpretation of the formation
mechanism is further complicated by use of the misleading
term "polymerization". It may, therefore, be worth examining the widely accepted definition of this term. Conventionally, "polymerization" means that the molecular
units (monomers) are linked together by the "polymerizing"
process. Specifically, rearrangement of the atoms or
elements that constitute the molecules of a monomer seldom
occurs during the polymerizing process. For instance, the
polymerization of styrene by conventional means is accomplished by the opening of a double bond without losing or
rearranging the atoms. Thus, the resulting polymer's name
is formed by using the term "poly" + the name of the monomer. In case of styrene, the polymer formed by the polymerization of styrene is named polystyrene. In other words,
polymerization in the accepted sense refers to molecular
polymerization, i.e., the chemical structure of a polymer
is denoted by the chemical structure of the monomer.

Strictly speaking, the ususal process of polymerization is not the same as the process that occurs in glow
discharge. In this context, polymer formation in a plasma
state takes place via a "nonpolymerizing" process. Thus,
polymer formation of this type may be characterized as
elemental or atomic polymerization in contrast to molecular
polymerization, which describes the conventional process.
The following discussion may illustrate the atomic nature
of polymer formation in a plasma state.

PLASMA-INDUCED AND PLASMA-STATE POLYMERIZATIONS

With regard to the preconceived concept of conventional (molecular) polymerization, the majority of monomers
that have been studied are conventional types, such as
styrene, ethylene, and vinyl chloride. Because the ionization of gas or vapor involves many highly energetic species,

which can trigger the polymerization of such monomers, and
as long as these monomers are used in plasma polymerization, it will continue to be difficult to distinguish
clearly between atomic and molecular polymerization. For
the sake of discussion, the conventional (molecular) type
of polymerization, which is triggered by a reactive
species created in an electric discharge, can be referred
to as "plasma-induced polymerization". The atomic process,
which occurs in a plasma state, can then be identified as
"plasma-state polymerization".

Plasma state polymerization can be represented by the
following mechanisms:

Initiation or Reinitiation

$$M_i \rightarrow M_i^*$$
$$M_k \rightarrow M_k^*$$

Propagation and Termination

$$M_i^* + M_k^* \rightarrow M_i-M_k$$
$$M_i^* + M_k \rightarrow M_i-M_k.$$

In these formulas, i and k are the numbers of repeating
units, i.e., i = k = 1 for the starting material, and M*
represents reactive species, which can be an ion of either
charge, an excited molecule, or a free radical produced
from M but not necessarily retaining the molecular structure of the starting material, i.e., M can be a fragment or
even an atom detached from the original starting material.
In this type of polymerization, the polymer is formed by
the repeated stepwise reaction described above.

Plasma-induced polymerization may be schematically
represented by a chain propagation mechanism as follows:

Propagation

$$M^* + M \rightarrow MM^*$$

$$M_1^* + M \rightarrow M_{i+1}^*$$

Termination

$$M_i^* + M_k^* \rightarrow M_i-M_k$$

It should be noted that plasma-induced polymerization does
not produce a gas phase by-product, because the process
proceeds via utilization of a polymerizable structure.
Overall, polymerization in a glow discharge consists of both
plasma-induced polymerization and plasma state polymerization. Which of these two mechanisms plays the predominant
role in the polymer formation in a glow discharce depends
not only on the chemical structure of the starting materials
but on the conditions of the discharge.

One of the most significant differences between plasma-
induced polymerization and plasma-state polymerization is
the fact that plasma-state polymerization produces gas
phase by-products, which do not become incorporated in the
polymer. This means that the components of the plasma
phase change as soon as plasma-state polymerization occurs.
Consequently, the influence of the product-gas plasma to the
entire system is an important factor.

EFFECT OF PRODUCT-GAS PLASMA

In an efficient glow discharge polymerization system,
a monomer tends to polymerize in the vicinity of the monomer inlet, and the range in which polymer deposition is observed is much narrower than the range in which glow discharge is observed. This situation is a reflection of the
fact that the plasma phase consists mainly of a product-gas
(e.g., H_2) and contains only a small amount of the monomer
or growing species. Because of this situation gas phase
analyses have failed to detect substantial amounts of
oligomers.

Because the gas phase changes from a monomer to product
gases as soon as glow discharge polymerization occurs, what
the product gas plasma does to the system, which consists of
the monomer, the intermediate species, a polymer, substrate
material, and the wall of the reaction vessel, becomes an
important factor. This factor depends on what kind and
how much of the product-gas or gases evolve during the process.

The major effects of the product-gas plasma may be seen
by examining the processes of 1) emission of photons and 2)
etching by chemical reaction. As far as the material balance in the system is concerned, these two processes can be
represented by ablation. Consequently, the entire glow dis-

charge polymerization can be schematically represented by
the Competitive Ablation and Polymer-formation (CAP) mechanism[1] shown in Figure 1.

Because most of the literature on glow discharge polymerization deals with hydrocarbons, which produce hydrogen
as the product gas, the effect of ablation does not appear
to be too great. Consequently, the complete neglect of
ablation does not make a significant difference in the overall picture of glow discharge polymerization; however, when
a fluorine-or oxygen-containing compound is used as the
starting material, the extent of ablation becomes predominant, and the amount of polymer formation depends entirely
on the quantity of gas produced.

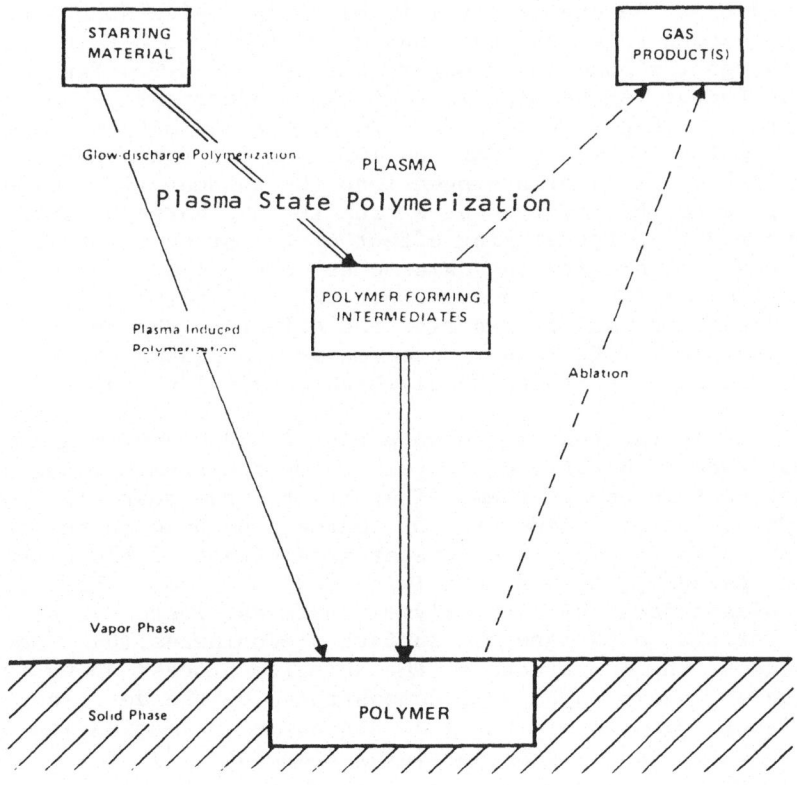

Fig. 1. Overall mechanism of glow discharge polymerization

   Perhaps the most dramatic demonstration of the effect
of ablation was recently given by Kay[2] in the glow discharge polymerization of CF_4. It had been thought that CF_4
was one of the very few organic compounds that does not
polymerize in a glow discharge. On the other hand, CF_4 has
been used as one of the most effective gases in plasma
etching. Kay observed that no polymer deposition occurs
under normal conditions in spite of the fact that the C-F
bonds are broken in the glow discharge, a fact which is con-
firmed by mass spectroscopic analysis of the gas phase.
However, when a small amount of hydrogen is introduced into
the discharge, the polymer is deposited. When the hydrogen
flow is stopped, the polymer deposit ablates.

   The situation observed in the above example may be
visualized by comparing the bond energies. It should be
noted that the energy level involved in a glow discharge is
high enough to break any bond.[3-4] That is, C-F is broken,
although its bond is stronger than C-H and C-C. The important factor is the stability of the product gas. The bond
energy for F-F is only 37 kcal/mol, whereas H-F is 135 kcal/
mol, which is higher than the 102 kcal/mol for a C-F bond.
The introduction of hydrogen into the monomer flow evident-
ly produces HF and removes F from the discharge system,
thus reducing the etching effect of the product-gas (F_2)
plasma and shifting the balance between polymerization and
ablation in favor of polymerization. Although the term
F_2 plasma is used to describe the effect of the detached F
in the plasma, F_2 is not detected in the plasma state, perhaps because of its extremely high reactivity. (5)

   It is interesting to note that F and O are two elements
that reduce the rate of polymer formation from compounds
that contain one of them. They are the two most electronegative of all elements. Of course, the bond energy itself is not a measure of the etching effect of the plasma.
For instance, the N-N bond is only 32 kcal/mol. However,
N_2 plasma does not etch polymer surfaces; instead, the incorporation of N into the surface predominates.(6)   Nevertheless, the importance of the ablation process shown in
Figure 1 seems to be well demonstrated by the deficient
polymer formation in the glow discharge polymerization of
CF_4, C_2F_6, and oxygen-containing compounds. (7)

   The polymer formation and properties of polymers

formed by glow discharge polymerization are controlled by
the balance among plasma-induced polymerization, plasma
state polymerization, and ablation, i.e., polymer formation
is a part of the CAP scheme shown in Figure 1. Because of
this type of glow discharge polymerization, the gas evolved
from the substrate also plays an important role, particularly at the early state of coating. It has been noticed
that the deposition of a polymer from the glow discharge of
styrene onto plyoxymethylene, oxygen-containing and plasma
labile substrate, is significantly different from deposition onto more stable substrates and is very dependent on
the conditions of the glow discharge.(8)

MOLECULAR POLYMERIZATION vs. ATOMIC POLYMERIZATION

Atomic polymerization is the principal element of the
CAP mechanism, but a similar result may be expected if one
conceives of a polymerizable precursor. The formation of
polymers from organic compounds, which do not polymerize
under normal conditions, may be postulated by assuming that
polymerizable precursors are formed in the plasma-state
and form polymers by conventional polymerization mechanisms.
For instance, the slower polymer deposition from ethylene
compared to that from acetylene has been attributed to the
slower process of forming acetylene, which is assumed to be
the precursor of glow discharge polymerization.(9) According to this concept, saturated hydrocarbons, such as
methane and ethane, will polymerize via plasma synthesis of
acetylene and subsequent (molecular) polymerization. The
frequently disputed subject of vapor phase polymerization
versus surface polymerization is discussed in the context
of precursor concept (although it is usually not mentioned
explicitly) because one cannot explain the polymerization
of saturated hydrocarbon, such as CH_4, without assuming
such a precursor.

There is no clear-cut answer to whether atomic polymerization theory or precursor theory explains the actual process of polymer formation in glow discharge better, because
there is no direct evidence to prove or disprove either.
All experimental data in glow discharge polymerization can
be interpreted only as circumstantial evidence so far as
the mechanism of polymer formation is concerned. Therefore,
the following discussion presented to support the concept
of atomic polymerization is not intended to disprove any

other concept but is intended to present a new way of recog-
nizing glow discharge polymerization. The most important
point is that any theory must cover all possible cases and
must satisfactorily explain not only one aspect of glow
discharge polymerization, e.g., polymer deposition rate,
but all other aspects, e.g., distribution of polymer deposition and change of polymer properties associated with
the distribution, in a consistent manner.

The following arguments used to support the concept of
atomic polymerization are not necessarily consistent with
others presented in the literature. This can be attributed
to the fact that glow discharge polymerization is highly
system-dependent, and the definition and expression of experimental conditions have been rather ambiguous. For instance, glow discharge polymerization has been customarily
carried out under arbitrarily chosen experimental conditions,
such as with a fixed discharge wattage and at a fixed flow
rate, but a satisfactory comparison of the glow discharge
polymerization of two different monomers cannot be made in
such a manner. Because the minimum discharge wattage necessary to effect the glow discharge polymerization of a
given monomer, e.g., ethylene, can differ greatly from
that for another, e.g., n-hexane, it becomes immediately
obvious that a comparison of these two monomers cannot be
made at a fixed discharge wattage. The data used in the
following discussion are based on the conditions of glow
discharge polymerization described in detail by Yasuda and
Hirotsu.(10)

Table I.   Comparison of Polymer Deposition Rates[a] for Vinyl
and Saturated Vinyl Compounds

Vinyl monomer		Saturated vinyl monomer	
Compound	$k \times 10^4$ (cm^{-2})	Compound	$k \times 10^4$ (cm^{-2})
4-Vinylpyridine	7.59	4-Ethylpyridine	4.72
α-Methylstyrene	5.33	Cumene	4.05
Styrene	5.65	Ethylbenzene	4.52
N-Vinylpyrrolidone	7.75	N-Ethylpyrrolidone	3.76
Acrylonitrile	5.71	Propionitrile	4.49
Vinylidene chloride	5.47	1,1'-Dichloroethane	2.98
Allylamine	2.86	n-Butylamine	2.52
Methyl acrylate	0.99	Methyl propionate	0.57

[a]Polymer deposition rate R in g/cm^2·min is given by $R = kF_w$; F_w is the weight basis flow rate
(g/min).

CORRELATION BETWEEN POLYMER DEPOSITION RATE AND CHEMICAL
STRUCTURE OF MONOMER

When polymer deposition rates are compared for various
pairs of monomers, which have similar chemical structures
with and without vinyl double bonds, the difference between
those with olefinic vinyl double bonds and those without
is very small[11] as shown in Table I. Furthermore, the
differences among various kinds of monomers are surprisingly small. In other words, nearly all hydrocarbon monomers
polymerize at rates that vary only within an order of magnitude. These two aspects indicate that no specific
structure (necessary for the precursor theory) is needed
for the glow discharge polymerization of organic compounds.

The dependence of polymer deposition rates on the molecular weight of monomers is another important aspect.
For instance, if one takes a homologous series of saturated
hydrocarbons, the probability of their forming precursor
structures, e.g., acetylene, decreases rapidly as the number of carbons increases, whereas the number of hydrogen
molecules evolved during glow discharge polymerization increases monotonously with the number of carbons[12] as
shown in Figure 2. This indicates that every hydrogen atom
has an equal probability for hydrogen abstraction. Therefore, if a precursor is formed first, the polymer deposition rate should decrease with an increase in the number
of carbons in a hydrocarbon molecule. Contrary to this
expectation, the polymer deposition rate increases with the
molecular weight of the monomer[7] as shown in Figure 3.
These correlations are in accordance with the atomic polymerization mechanism.

Characteristics of atomic polymerization are directly
seen in the incorporation of gases or vapors, such as N_2,
CO, and H_2O, by polymers formed in glow discharge, when
these gases or vapors are mixed with the vapor of an organic compound.[13,14]

If the basic step of forming a polymer is molecular
polymerization of the precursor species created in a plasma,
the incorporation of such gases or vapors cannot be explained. This gas incorporation in glow discharge polymerization is also an indication of the atomic nature of the

polymer formation process, i.e., gases and vapors provide
atoms but are not incorated as molecules.

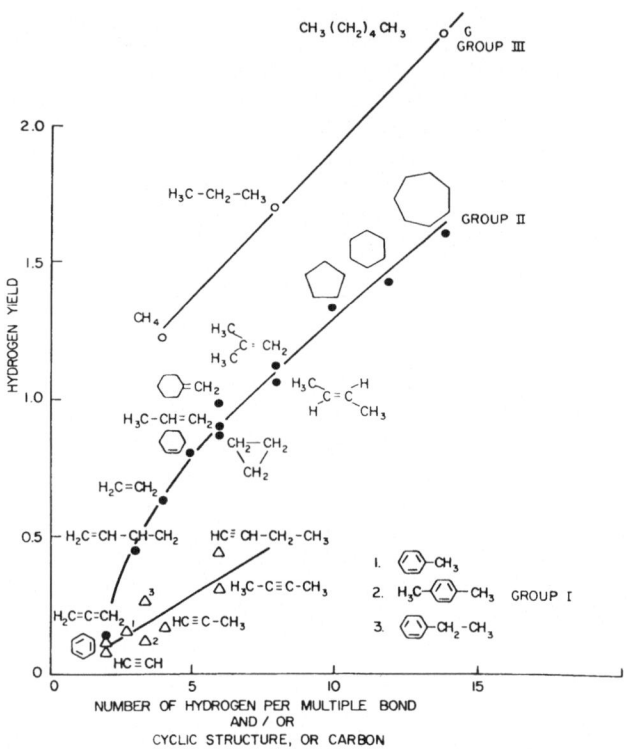

Fig. 2.   Hydrogen yield of monomers of plasma polymerization

TRAPPED FREE-RADICALS IN POLYMERS

One of the most significant features of glow discharge
polymerization is that large amounts of free-radicals are
trapped in the polymer.(15-16) Although the amounts of
trapped free-radicals vary with the types of monomers and
conditions of glow discharge polymerization, it is safe to
consider that glow discharge polymers contain trapped free
radicals. An explanation of how these free radicals are
formed could provide important information for how polymeric
materials are formed in the glow discharge or organic
compounds.

Obviously, molecular polymerization, e.g., free radical polymerization of vinyl monomers, does not yield polymers with trapped free-radicals. Therefore, if one considers molecular polymerization of precursors as the main polymer forming mechanism, it is necessary to take into account separate mechanisms for creating free radicals in polymers. An easily conceived mechanism is free radical

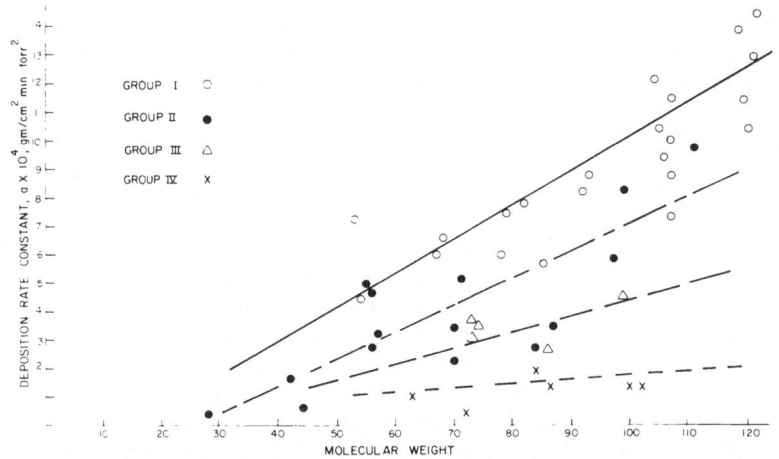

Fig. 3. Dependence of polymer deposition rate on molecular weight of monomer. Group I - Triple-bond-containing, aromatic, and hetearomatic compounds. Group II - Double-bond-containing and cyclic compounds. Group III - Compounds without above mentioned structures. Group IV - Oxygen-containing compounds.

formation by radiation, because glow discharge can be considered as a kind of radiation process. There are many energetic species, such as electrons, ions, excited molecules, free radicals and photons, in glow discharge, consequently such a hypothesis is not at all unreasonable. Therefore, the trapping of free radicals in a glow discharge polymer as a function of the chemical structure of monomers should be examined.

First, there is a definite correlation between the spin concentration measured by ESR and the chemical structure of monomers. Glow discharge polymers of monomers that

contain triple bond, aromatic, and hetearomatic rings and
$-C\equiv N$ produce the highest level of free spins. Hydrocarbons
can be grouped into three major types based upon their behavior in glow discharge polymerization. These can be
tentatively referred to as Group I, II, and III monomers.
Group I monomers are those mentioned above. They polymerize with evolution of the least amount of hydrogen and
contain the highest level of free spins among polymers.
Group II monomers are compounds containing an olefinic
double bond and/or a cyclic structure. Group III monomers
are saturated compounds, which do not contain the structures mentioned in Groups I and II. Group III monomers
polymerize with evolution of the highest level of hydrogen and contain the least amount of free spins. Group II
monomers lie in between these two extremes, i.e., they
evolve a moderate amount of hydrogen and exhibit an intermediate level of free spin concentration.

Because the energy level of the electrons and ions are
much lower compared to those involved in other radiation
processes, such as in γ-rays and high energy electron
beams, it is thought that these energetic species have no
penetrating power. This agrees with the ESR data obtained
with glow discharge polymers applied to glass substrates.
(15) The free spin concentration of the samples can be
further divided into the free spin concentration in the
glow discharge polymers and in the substrate glass tubes.

The free-spin concentration in the glass can be attributed to ultraviolet irradiation in the glow discharge
polymerization system. If trapped free-radicals are formed
by the irradiation of the formed polymer by molecular
polymerization of the plasma synthesized precursor and,
consequently, imparting no free-radicals in the polymer,
the level of ultraviolet irradiation manifested by the
free-spin concentration in the glass should be proportional to the free-spin concentration in glow discharge
polymer. In other words, the highest free spins in glass
should be obtained by the glow discharge polymerization of
Group I monomers. Contrary to this expectation, the highest level of ultraviolet emission is associated with Group
III monomers, which yield the least amount of free spins
in the polymer as shown in Table II. This trend is in
accord with the intensity of the glow observed in the glow
discharge polymerization of monomers. Specifically, the
glow observed for Group I is very weak, whereas the one

observed for Group III is the most intense. In other words,
the level of ultraviolet emission is proportional to the
evolution of hydrogen in the glow discharge polymerization
system. Consequently, the hypothesis that trapped free
radicals are formed by irradiating polymers does not explain the forming of trapped free-radicals in glow discharge polymers. Thus, the precursor theory cannot explain the presence of large amounts of free-radicals in
glow discharge polymers. That means that a radically new
approach is needed to explain the formation of polymers in
glow discharge.

TABLE II

ESR Spin Concentration in Plasma Polymers
and Glass Substrates

Monomer	$C_s \times 10^{19}$, spins/cm³		$C_g \times 10^{-15}$, spins/cm² *	
	Continuous	Pulsed	Continuous	Pulsed
C_2H_2	8.6	15.6	0	0
C_6H_6	3.2	1.6	0	0
C_6F_6	7.4	5.4	0	0
Styrene	3.8	0.54	0	0
C_2H_4	1.36	14.5	4.0	0.85
C_2F_4	13.0	8.4	11.2	1.8
Cyclohexane	0.84	0	1.1	0
Ethylene oxide	0.75	0.5	6.6	1.6
Acrylic acid	0.76	1.85	4.4	0
Propionic acid	1.0	1.0	6.3	1.6
Vinyl acetate	0.42	0.33	6.1	1.8
Methyl acrylate	0.31	0.15	6.4	1.5
Hexamethyldisilane	0.5	0.24	0	0
Tetramethyldisiloxane	0.49	0.05	0	0
Hexamethyldisiloxane	0.21	0	0	0
Divinyltetramethyldisiloxane	0.15	0.05	0	0

*Total spins divided by surface area of glass.

SPUTTER-COATING vs. GLOW DISCHARGE POLYMERIZATION

The atomic polymerization so far discussed can be
characterized by the fact that a monomer used in glow discharge polymerization provides a source of elements but
does not act as a monomer in a molecular sense. Similar
phenomena can be seen in the sputtering of metals. Therefore, it may be worth comparing the sputtering process with
the glow discharge polymerization of organic compounds.

The sputtering of metal is usually conducted by d.c. glow discharge of argon and using the target metal as a cathode. The accelerated ions hit the target metal on the cathode and transfer the energy to the target atoms. The atoms, which receive energy from the impinging ions, shake up the structure and strike neighboring atoms. This energy transferring process ends up with some atoms leaving the solid phase and sputtering away into the gas phase. If a substrate material is placed in the gas phase, the sputtered material is deposited onto the substrate. This is one of the methods used to metallize polymer surfaces. Because the shake-up process occurs at the level of the atoms, alloys, such as stainless steel, can be used for sputter-coating, whereas metallization by evaporation cannot be used for alloys.

Glow discharge polymerization has many aspects in common with sputter-coating. Both processes are initiated by ionization of a gas (or vapors as in glow discharge polymerization). The major differences are the location of target material and the substrate. In the sputter-coating process, the target material is the cathode surface, and the substrate for coating is placed outside of the glow discharge. To avoid the effects of plasma, special efforts are often made to contain the glow discharge. For example, in a magnetron discharge, the superimposed magnetic field is used in the glow discharge polymerization, the target material is the monomer in a vapor phase, and the substrate is placed inside the glow discharge. The ionization process or collision of ions or electrons with molecules of the monomer "sputters" the fragments of the monomer molecules and deposits them on the substrate. Therefore, it may be assumed that both sputter-coating and glow discharge polymerization are atomic processes.

One of the unique characteristics of polymers formed by glow discharge polymerization is their strong tendency to exhibit internal (expansive) stress in the polymer layer during the process of deposition. When a thin layer of glow discharge polymer is deposited on a thin polymer film, the coated film tends to curl as a result of the internal stress in the coating. By knowing the thickness of both layers and the Young's modulus of the substrate film, the curling force and the internal stress can be calculated from the radius of the curled sample.(17) Typical cases are shown in Figure 4 and Table III.

This high level of internal stress sometimes causes self-distinction of the coating when too thick a layer is deposited. An important aspect is that the build-up of internal stress, manifested by the curling of the coated substrate, occurs during the process of polymer deposition. This aspect, therefore, should be directly related to the mechanism of polymer formation. Another possible cause of the internal stress is the absorption of oxygen by trapped free-radicals and subsequent absorption of water vapor by the oxygen containing functions. These two phenomena, which occur in general cases, will cause expansion or swelling of the polymer deposit. However, it has been found that there is no direct correlation between the free spin concentration observed in ESR and the reported internal stress. (14) This indicates that the major portion of the build-up of internal

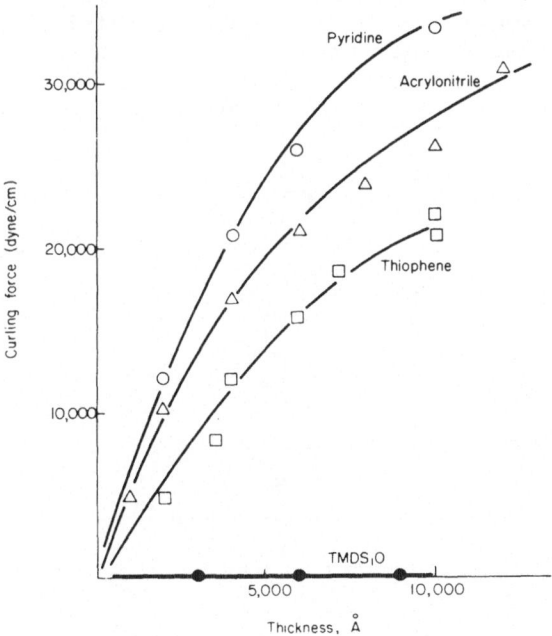

Fig. 4. Curling force $\sigma_s d$, which is the product of the internal stress (σ_s) and the thickness (d) of the plasma-deposited layer, is plotted vs d. The three curves correspond to layers obtained by plasma polymerization of the indicated monomers.

stress occurs during the process of polymer deposition.   This
aspect is very similar to the stress build-up that takes
place during the deposition of sputtered metal, thus indicating that the mechanism is alike in both cases.

Polymers formed by glow discharge polymerization have
much greater densities than corresponding conventional
polymers.(18)   For instance, the density of the glow discharge polymer of ethylene was found to be as high as 1.3,
but the polymer has no crystallinity.   The presence of high
density and high expansive internal stress in the absence
of crystallinity can be explained by the continuous impinging
process in atomic polymerization, whereas the molecular
polymerization mechanism offers no such reasonable explanation.

Table III
Internal Stress of Plasma Polymer
(thickness = 4000 Å; reaction conditions;
30 μm Hg monomer pressure and 80W power)

Monomer	Structure	σ_s, dynes/cm^2
Thiophene		2.7×10^8
Pyridine		5.2×10^8
Acrylonitrile	$CH_2{=}CH{-}C{\equiv}N$	4.3×10^8
Furan		7.0×10^8
Styrene	$CH_2{=}CH$	4.3×10^8
Acetylene	$CH{\equiv}CH$	3.8×10^8
2-Methyloxazoline		0
Tetramethyldisiloxane	$H{-}\underset{Me}{\overset{Me}{Si}}{-}O{-}\underset{Me}{\overset{Me}{Si}}{-}H$	0

STRUCTURAL DIFFERENCE OF MONOMER AND POLYMER

The aspect of atomic polymerization can be clearly seen
in the molecular structural difference of monomers and polymers, but with the glow discharge polymers of hydrocarbons,
without appropriate analytical tools, it is difficult to

demonstrate the difference in a (semi) quantitative manner,
although such tests as elemental and infrared spectra
analyses generally show that glow discharge polymers are
quite different from corresponding conventional polymers.
Specific analysis is hampered by the characteristic insolubility of most glow discharge polymers. However, the
use of Electron Spectroscopy for Chemical Analysis (ESCA)
to study the glow discharge polymers of perfluorocarbons
provides a unique opportunity to overcome these difficulties.
The results of ESCA studies of glow discharge polymers of
tetrafluoroethylene, $CF_2=CF_2$, are reviewed below in view
of atomic polymerization.

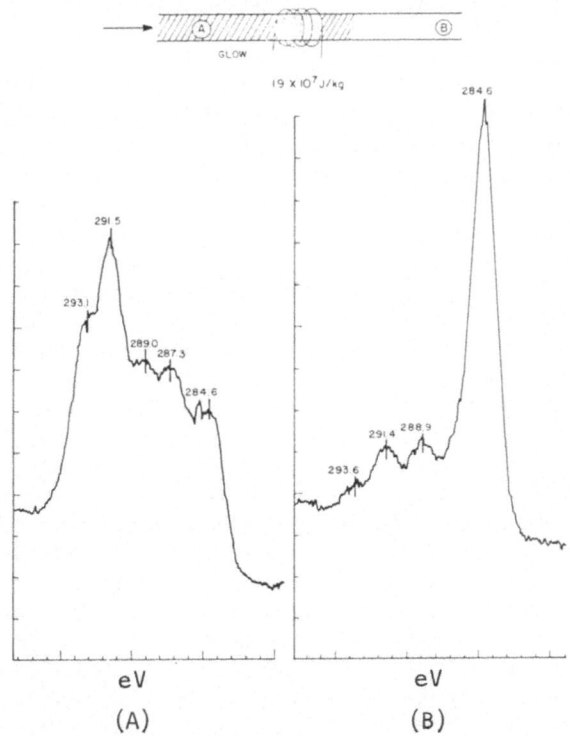

Fig. 5. Dependence of the ESCA Cl_s peaks of glow discharge
         polymers of tetrafluoroethylene on discharge con-
         ditions and the location of polymer deposition.
         Polymer deposit occurred at two locations (a) before
         the rf coil, and (b) after rf coil. Discharge power
         level is 1.9 x 10^7 Joules/kg.

Because of the strong electron negativity of fluorine atoms, the binding energy of the core level electron of carbon is shifted enough so that the amount of shift can be measured by the ESCA Cl_s spectrum. Consequently, ESCA can distinguish carbons that have one, two, or three fluorine atoms attached. Conventional polytetrafluoroethylene (Teflon) shows a singlet Cl_s peak, which corresponds to $-CF_2-$ at 291.5 e.v. This peak is shifted from the Cl_s peak of normal carbons (bound to H or C) at 284.5 e.v.

If a polymer is formed by plasma-induced polymerization in a glow discharge, the Cl_s peak should be a singlet at 291.5 e.v. as it is in the case of polytetrafluoroethylene. If a polymer is formed via a triple bond containing a precursor as it is proposed for the glow discharge polymerization of hydrocarbons, the Cl_s peak should be at a lower binding energy level (289 e.v.) than that for $-CF_2-$, because the formation of such a structure requires the abstraction of two fluorine atoms from two adjacent carbons to yield polymers with less fluorines. In the most typical cases, however, the Cl_s peak of the glow discharge polymer of tetrafluoroethylene contains a considerable amount of $-CF_3$, $-CF_2-$, and intermediate peaks in between $-CF_2-$ and C as shown in Figure 5. This strongly indicates that polymers are formed by neither molecular polymerization of the monomer nor (molecular) polymerization of the plasma-synthesized precursor. This situation is exactly what is to be expected of the atomic polymerization mechanism described earlier for plasma-state polymerization. (Namely M* can be a fragment of a molecule including a single atom).

Other evidence, which supports the concept of atomic polymerization, may be seen under certain conditions in the codeposition of aluminum (used as a substrate) in the glow discharge polymer of tetrafluoroethylene. When an excessive discharge power is used for the glow discharge polymerization of tetrafluoroethylene, the fluorine detachment and the consequent ablation prevail in the polymer formation.(1) Under such a condition, the ESCA Cl_s peak indicates very little $-CF_3$, $-CF_2-$, and the major peak becomes a broadened peak around 285.5 e.v. When this happens, the polymer deposition rate decreases drastically, and the ESCA spectrum shows the presence of Al as shown in Figure 5. The ESCA Al 2_s peak observed in this case is not identical to that of the Al foil used as the substrate, and the Fl_s peak also shows a

conspicuous doublet, indicating that aluminum fluoride is
formed and co-deposited in the glow discharge polymer.

CONCLUDING REMARKS

Atomic polymerization is not polymerization in the conventional sense, because the molecular structure of the
monomers is not retained in the polymer. For instance, the
glow discharge polymers of acetylene and benzene are very
much alike. Their copolymerization characteristics are
nearly identical. When N_2 and H_2O are added to benzene or
acetylene, considering that one molecule of benzene is
equivalent to three molecules of acetylene in glow discharge
polymerization, nearly identical polymers are formed.(19)
Yet, neither the glow discharge polymer of acetylene or
benzene is quite the same as the glow discharge polymer of
either methane or ethane. Therefore, the chemical structures
of the monomers do play an important role in glow discharge
polymerization, although the original structures or their
derivatives may not be retained in the polymer structures.
In this sense, absolute atomic polymerization can be seen
in carbon films deposited from hydrocarbons when subjected
to glow discharge. As shown in Figure 1, the overall glow
discharge polymerization generally occurs by simultaneous
atomic and molecular polymerizations.

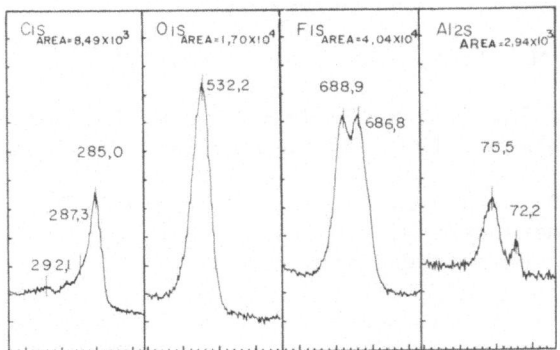

Fig.   6.   ESCA spectrum of glow discharge polymer of tetra-
            fluoethylene polymerized at high energy input
            (W/FM).

If one accepts the concept of atomic polymerization as a mechanism of polymer formation in the glow discharge of organic compounds, little difference can be seen between the glow discharge polymerization and glow discharge deposition of inorganic materials, such as carbon, SN_x and Si. Thus, the true significance of glow discharge polymerization will be found in the preparation of entirely new materials, particularly hybrids of organic and inorganic materials rather than a method of polymerization in the conventional sense, which yields ill-defined polymers.

ABSTRACT

Polymeric materials are formed from a number of organic compounds under conditions of low temperature plasma (ionized vapor). The process is generally called "plasma polymerization" or "glow discharge polymerization"; however, the formation of polymeric materials in plasma polymerization is so much different from the generally accepted mechanism of polymerization that the term "polymerization" may not represent the actual process. The differences between the conventional (molecular) polymerization and polymer formation in the plasma state, which can be considered "atomic" or "elemental" polymerization, are explained to provide an understanding of the mechanism and significance of the latter.

LITERATURE CITED

1. Yasuda, H. and Hsu, T., Surf. Sci., 26, 232, 1978.

2. Kay, E., Intl. Round Table on Plasma Polymerization and Treatment, IUPAC Symp. on Plasma Chemistry, 1977.

3. Clark, D. T. and Dilks, A., "Characterization of Metal on Polymer Surface," in Polymer Surfaces, Vol. 2, Academic Press, New York, 1977.

4. Wehner, G. K. and Anderson, G. S. in Handbook of Thin Film Technology, Maissel, L. I. and Glang, R., Eds., McGraw-Hill, New York, 1970.

5. Smolinsky, G., Bell Telephone Laboratories, private communication, August 1977.

6. Yasuda, H. Marsh, H. C., Brandt, E. S. and Reilley, C.
    N., J. Polym. Sci., Polym. Chem. Ed., 15, 991, 1977.

7. Yasuda, H., Intl. Round Table on Plasma Polymerization
    and Treatment, IUPAC Symp. on Plasma Chemistry, 1977.

8. Yasuda, H., Lamaze, C. E. and Sakaoku, K., J. Appl.
    Polym. Sci., 17, 137, 1973.

9. Kobayashi, H., Shen, M. and Bell, A. T., J. Macromol.
    Sci. Chem., A8 (2), 373, 1974.

10. Yasuda, H. and Hirotsu, T., J. Polym. Sci., Polym. Chem.
    Ed., 16, 743, 1978.

11. Yasuda, H., J. Macromol. Sci. Chem., A10 (3), 383, 1976.

12. Yasuda, H., Bumgarner, M. O. and Hillman, J. J., J.
    Appl. Polym. Sci., 19, 531, 1975.

13. Yasuda, H. Marsh, H. C., Bumgarner, M. O. and Morosoff,
    N., J. Appl. Polym. Sci., 19, 2845, 1975.

14. Yasuda, H. and Hirotsu, T., J. Polym. Sci., Polym. Chem.
    Ed., 15, 2749, 1977.

15. Morosoff, N., Crist, B., Bumgarner, M., Hsu, T. and
    Yasuda, H., Macromol. Sci., A-10 (3), 451, 1976.

16. Millard, M., in Techniques and Applications of Plasma
    Chemistry, Hollahan, J. R. and Bell, A. T., Eds., John
    Wiley, New York, 1974.

17. Yasuda, H., Hirotsu, T. and Olf, H. G., J. Appl. Polym.
    Sci., 21, 3179, 1977.

18. Knickmeyer, W. W., Peace, B. W. and Mayhan, K. G., J.
    Appl. Polym. Sci., 18, 301, 1974.

19. Yasuda, H., Marsh, H. C. and Tsai, J., J. Appl. Polym.
    Sci., 19, 2157, 1975.

PART II
CHARACTERIZATION

FOURIER TRANSFORM INFRARED SPECTROSCOPY OF THE POLYMERIC AMORPHOUS PHASE

Jack L. Koenig and Donald Kormos

Case Western Reserve University

Cleveland, Ohio 44106

INTRODUCTION

The characterization of the crystalline phase of semicrystalline polymers in terms of chain order and morphology has made considerable progress using diffraction and microscopic techniques. Characterization of the amorphous phase in terms of chain structure and conformation has been limited. Consequently, our knowledge of the amorphous phase for many polymeric systems is lacking. This is highly undesirable since many of the processes which are of interest (crystallization, diffusion, swelling, etc.) are initiated in the amorphous phase.

Vibrational spectroscopy potentially has sufficient sensitivity and specificity to detect chain conformations in solid polymer samples. An analogy can be drawn between the composite infrared spectrum arising from n-compound mixtures of simple organic solvents and the composite spectrum due to differing chain conformations in amorphous polymer glasses. Each mixture spectrum is a linear combination of a set of spectra assignable individually to the components of the mixture. In our study of polymer conformations, we will need techniques to analyze sets of mixture spectra for the number of basic spectra (e.g. number of conformations) and to derive the individual component spectra themselves. Furthermore, a method of internal calibration of the component spectra is necessary for measuring amounts of polymer conformations since the spectra of the pure component for any given conformation cannot be externally measured.

Factor analysis affords a method of determining the number of components in a series of mixtures. This technique has been recently applied to infrared spectroscopy (1). The absorbance ratio method already has had application in determining the component spectra (2, 3). A recent refinement of the ratio method suggests a method of internal calibration which allows the quantitative determination of the amounts of each component in the spectra of the amorphous phase (4).

We will illustrate the integrated use of the above techniques to study the amorphous and semicrystalline phases of poly(ethylene terephthalate) (PET). Factor analysis will be used to determine the number of spectrally identifiable components in a series of PET films annealed to different levels. The polymer has two rotational conformations assignable to glycol linkages; the most stable being the extended <u>trans</u> conformation (dihedral angle of 180°) and slightly higher in energy, the <u>gauche</u> conformation with the methylene units rotated out of the plane of the ring carbon atoms (dihedral angle of 60°). The glycol conformation has been shown by x-ray experiments to be exclusively <u>trans</u> in the crystalline phase of the polymer. The absorbance ratio method will be utilized to obtain the appropriate number of component spectra. A self-calibrating system will be used since external standard spectra are unobtainable for any real polymer system. With representative conformational spectra, a least-squares curve fitting procedure can be applied to analytically determine the concentration of the conformations in the PET films or arbitrary PET samples (5).

All of the instrumental work of this paper was done using a Fourier transform infrared spectrometer. The excellent spectral registration, precision, and data processing capability of the instrument were essential to the types of studies outlined.

THEORETICAL BACKGROUND

Factor analysis is a mathematical procedure for determining the number of linearly independent components in a set of mixtures. This procedure has been applied to the study of the infrared spectra of mixtures (1). Presented with a series of polymer films each having a different thermal history, factor analysis of the infrared spectra of these films should provide information pertaining to the number of spectrally isolable conformational components contributing to the spectra.

Consider a matrix, A, dimensioned m by r where m represents a given mixture spectrum and r the absorbance value for the spectrum m for each spectral element (in cm^{-1}). Simply, it is a listing of each of the mixture spectra by column. We are interested in obtaining the rank of the A matrix. The rank will be equal to the number of linearly independent basis spectra which make up the series of mixture spectra. The task of determining the rank of A is accomplished by first multiplying A by A^T (A transpose) yielding a matrix, C, termed the covariance matrix. The dimensions of C are condensed to m by m. The rank of C is equivalent to the rank of A. Diagonalizing C allows calculation of the eigen values and eigenvectors for the matrix. The rank of C is equal to the number of non-zero eigenvalues of C, the covariance matrix, which also indicates the number of component spectra in the initial set of mixture spectra. Experimentally, some degree of discernment is necessary in selecting non-zero eigenvalues since random spectral noise contributes small positive eigenvalues.

An absorbance ratio method has been applied to mixtures of simple organic molecules and also to polymers (2,3) to determine the component spectra without separation or isolation. The method can also be applied to determine the relative concentrations of the components without external calibration (4).

For binary mixtures, two different mixture spectra have absorbances (assuming constant path length) which can be represented by equations (1) and (2). A_m and A_m' represent the measured absorbances for the two components in mixtures m and m'

$$A_m = a_1 c_1 + a_2 c_2 \tag{1}$$

$$A_m' = a_1 c_1' + a_2 c_2' \tag{2}$$

$$\frac{A_m'}{A_m} = \frac{a_1 c_1' + a_2 c_2'}{a_1 c_1 + a_2 c_2} \tag{3}$$

$$\left(\frac{A_m'}{A_m}\right)_{r_1} = \left(\frac{a_1 c_1'}{a_1 c_1}\right)_{r_1} = \frac{c_1'}{c_1} = R_{r_1} \tag{4}$$

$$\left(\frac{A_m'}{A_m}\right)_{r_2} = \left(\frac{a_2 c_2'}{a_2 c_2}\right)_{r_2} = \frac{c_2'}{c_2} = R_{r_2} \tag{5}$$

mixture spectra each composed of a concentration factor, c, times the pure component absorbance, a, in the most general sense. Equation (3) defines the ratio spectrum of A_m and A_m'. In regions of the ratio spectrum where the spectral contribution of a_2 is absent (frequency equal to r_1), the ratio is simplified to the ratio of the concentration factors for the two mixtures as detailed by equation (4). The ratio coefficient, R_{r_1}, is therefore defined. Similarly, in regions of the ratio spectrum where a_1 is absent, the ratio coefficient R_{r_2} is experimentally obtained. As shown in equation (5). Since the concentration factors are defined as volume fractions they must add to unity

$$c_1 + c_2 = c_1' + c_2' = 1 \tag{6}$$

we can solve for the concentration coefficients as a function of R_{r_1} and R_{r_2}:

$$c_1 = \frac{1 - R_{r_2}}{R_{r_1} - R_{r_2}} \qquad c_2 = \frac{R_{r_1} - 1}{R_{r_1} - R_{r_2}} \qquad c_1' = R_{r_1} c_1 \qquad c_2' = R_{r_2} c_2 \tag{7}$$

Since A_m and A_m are measured, the unmixed pure component spectra A_1 and A_2 can be obtained

$$A_1 = \left(\frac{1}{1-R_{r_2}}\right)A_m' - \left(\frac{R_{r_2}}{1-R_{r_2}}\right)A_m \tag{8}$$

$$A_2 = \left(\frac{1}{1-R_{r_1}}\right)A_m' - \left(\frac{R_{r_2}}{1-R_{r_1}}\right)A_m \tag{9}$$

It is important to note that A_1 and A_2 are properly scaled to one another and as such can be used in least-squares fitting routines. This represents the internal calibration of the absorbance ratio technique.

Adherence to the Beer-Lambert law, which implies the linear addition of component spectra in a mixture spectrum, is required by the ratio method. Band overlap can disturb the determination of ratio coefficients as does background baseline intensity.

Once the component spectra of a mixture have been determined and properly scaled, perhaps via the ratio method, other mixtures composed of identical components can be analyzed using the method of least-squares curve-fitting. A program to do so has been developed in our laboratory (5). The fitting equation employed is shown in equation (10). N is the number

$$\sum_{i=1}^{N} R_{i,k} = \sum_{j=1}^{M} x \, (\sum_{i=1}^{N} W_i R_{i,j} \, R_{i,k}) X_j \tag{10}$$

$$S_i = \sum_{j=1}^{M} X_j R_{i,j} \tag{11}$$

of spectral elements (data points) in each spectrum and M the number of basis spectra used in the fitting procedure. $R_{i,k}$ represents the absorbance data for the i-th spectral element of the k-th basis spectrum. S_i is the data for the spectral range of the mixture spectrum. W_i is a weighting factor equal to $1/S_i$ (statistical weighting). X_j is the number which when multiplied by the appropriate basis spectrum and summed for all components best fits the experimental mixture spectrum. When using calibrated component spectra, it is from the computer determined X_j values that volume fractions of the components in the mixture can be obtained.

EXPERIMENTAL

Two types of polymer samples were used in the experiments that will follow. Solvent-cast amorphous poly(ethylene terephthalate) (PET) film of approximately 3 mils in thickness was obtained from E. I. du Pont de Nemours and Co. This film was of sufficient thickness to record the weak carbon-hydrogen stretching region for the polymer. To accurately record the fingerprint region of PET, solution-cast films were made using a solution of 1.5 grams of 1.0 mil commercial Mylar film (E. I. du Pont de Nemours and Co.) in 10 milliliters of an 80/20 (by volume) mixture of chloroform and trifluoroacetic acid (both spectral grade). A thin layer of the solution was spread uniformly on the surface of a clean glass slide. The solvent was allowed to evaporate in air and the films carefully floated off the glass surface in distilled water. The films were dried and stored under vacuum in a dessicator for at least one week at room temperature.

Spectra were obtained on a Digilab FTS-14 Fourier transform infrared spectrometer using double precision software. All spectra were obtained at 2 cm^{-1} resolution and averaged over a minimum of 200 scans. Data analysis programs for factor analysis and least squares curve-fitting were written in FORTRAN and were compiled to include double precision real variables and functions. All spectra were kept below 1.00 absorbance unit to assure the validity of the Beer-Lambert relation. Spectra shown are in absorbance units.

RESULTS

Figure 1 illustrates the effect of annealing an amorphous PET film in the carbon-hydrogen stretching region. The spectra shown include the amorphous spectrum and the spectrum of the same sample after annealing at 160°C. for 2 hours. A shoulder band at 2909 cm^{-1} becomes much more pronounced as does a band at 3015 cm^{-1}. Increases in the amount of the <u>trans</u> conformer were suggested due to either changes in the amorphous state and/or crystallization.

Detailed in Figure 2 is an application of factor analysis to a set of 8 spectra taken of annealed PET that covers the range of amorphous to annealed at 200°C for 3 hrs. The figure shows a plot of log eigenvalue versus components. Two eigenvalues are

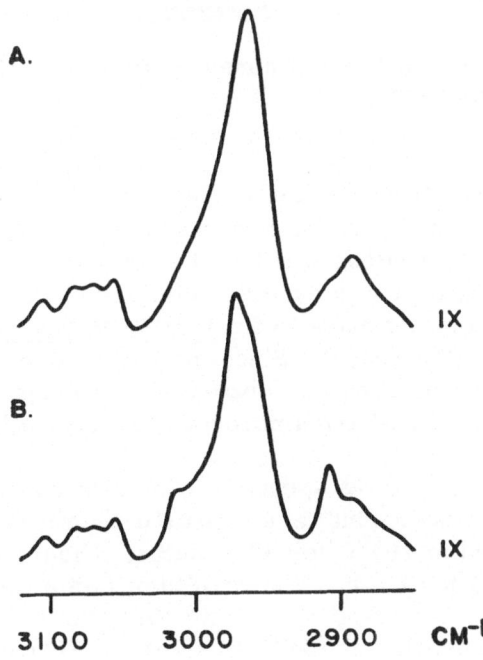

Figure 1. PET C-H stretching region: (A) unannealed film; (B) film annealed at 160°C for 2 hrs.

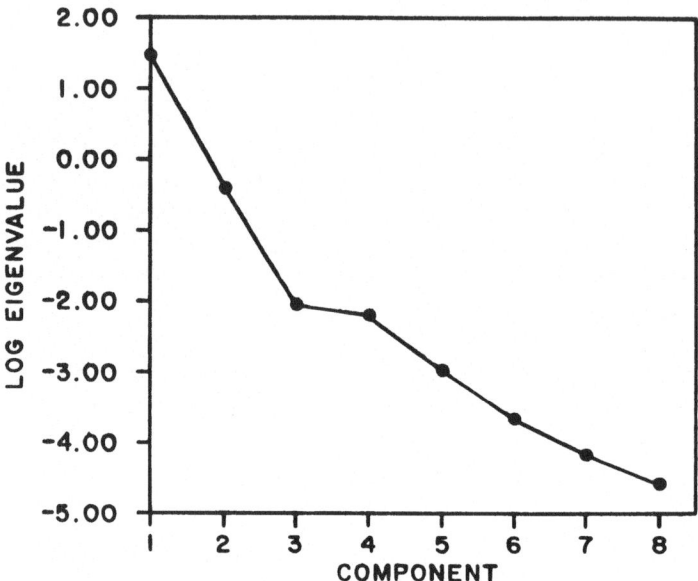

Figure 2.   Factor analysis of annealed PET films (3 mil)
               3120-2850 cm^{-1}.

clearly separated from the test of the values. By analogy to the
work of Koenig and Antoon on factor analysis of infrared spectra
(1), these two eigenvalues will be taken as non-zero and the others
assumed to be zero due to experimental error. Assignment of
the two components is made to the trans and gauche rotational
isomers of the PET repeat. Since crystallization under the
annealing conditions used was observed, the trans component
must be present in both the amorphous and crystalline domains.

    Figure 3 details the spectrum of amorphous PET in the
1700-800 cm^{-1} fingerprint region and the spectrum of the same
sample annealed at 104°C for 92 minutes. Clear evidence of
spectral change is shown. The results of factor analysis for a
series of 13 spectra taken of a single PET thin film after
cummulative annealing at 104°C is shown in Figure 4. The
annealing treatments covered the amorphous to semicrystalline
range. The analysis was carried out over the entire 1700-800
cm^{-1} range. Two eigenvalues are clearly distinct from the rest

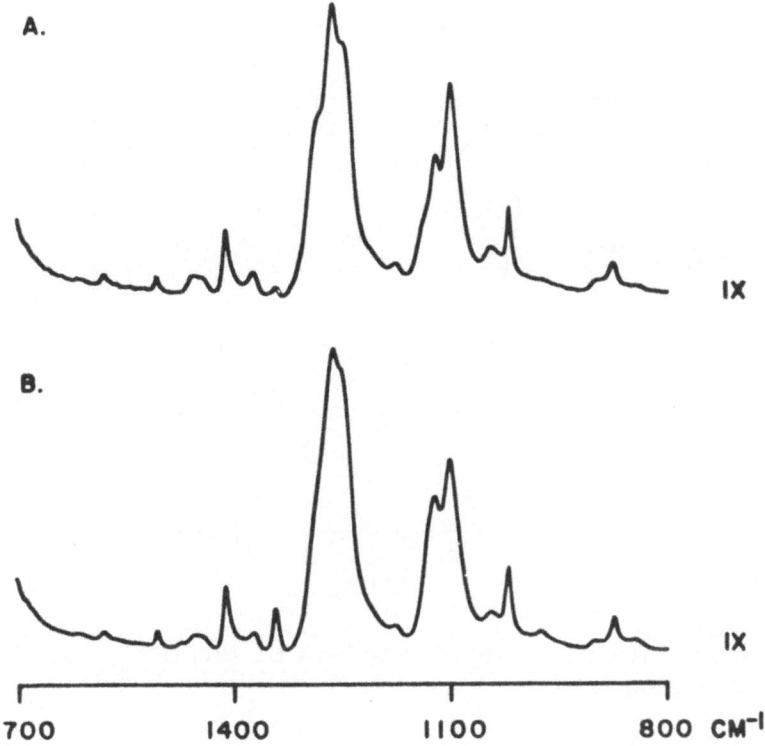

Figure 3.   PET fingerprint region: (A) unannealed thin film;
            (B) thin film annealed at 104°C for 92 min.

in the plot of log eigenvalue versus components. These values
are assigned as the non-zero eigenvalues and a 2 component
mixture system is found. This concurs with the results of the
C-H stretching region. Assignment of the mixture components
is to the gauche isomer in the amorphous phase and the trans
isomer present in both the amorphous and crystalline phase.

    Factor analysis has shown that for samples ranging from
amorphous to semicrystalline, the spectra of PET in the C-H
stretching and fingerprint regions can be modeled as a two
component system the trans and gauche isomers reflect the two
components.

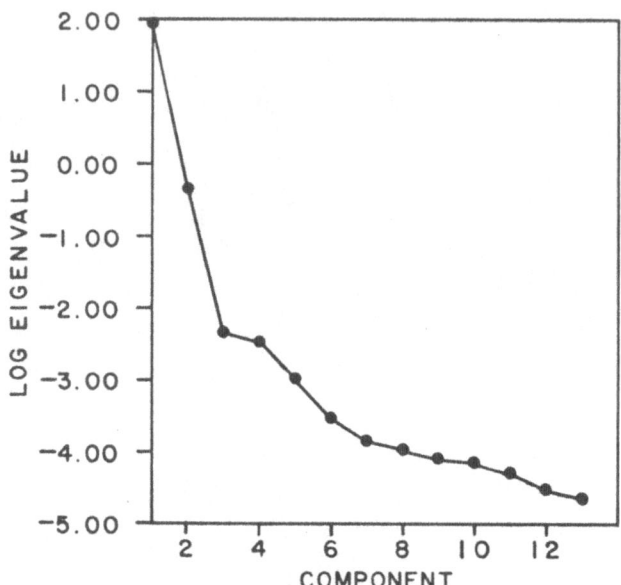

Figure 4.   Factor analysis of annealed PET thin films, 1700–
           800 cm^{-1}.

       With the number of components determined, the ratio method
can now be applied to the binary mixture of conformational
isomers in PET. We seek information regarding the spectra of
the two components as well as a measure of the amounts of each
component in the samples investigated.

       For the C–H stretching region, a ratio plot of an annealed
film divided by an unannealed amorphous film is shown in Figure
5. The ratio coefficients R_{r1} and R_{r2} are determined to be
2. 170 and 0. 626 respectively. Using this information and applying
the ratio technique, the concentration of the trans isomer for the
unannealed sample was found to be 47. 4 %, the remainder being
gauche. The film annealed for 2 hrs. at 160°C was found to be
75. 8% trans with 24. 2% gauche. Figure 6 details the deconvoluted
and internally scaled gauche and trans spectra. The trans
spectrum shows characteristic bands at 2909, 2972, and 3015 cm^{-1}

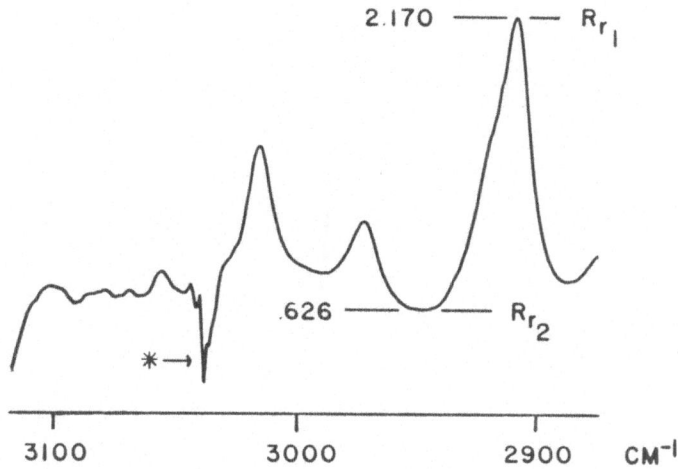

Figure 5. Ratio spectrum for 3120–2850 cm^{-1} region;
           annealed at 160°C for 2 hrs divided by unannealed.
           *Indicates artifact peak.

as well as C–H ring stretching modes at higher frequencies.
The trans results agree with the theoretical vibrational analysis
for crystalline PET by Boerio et al. (6).

A ratio plot of two film spectra after different annealing
histories for the appropriate bands in the fingerprint region can
determine gauche and trans spectra utilizing the ratio method.
For a thin film annealed for 32 min. at 104°C divided by one
annealed for 1 min. at 104°C, the gauche and trans spectra
shown in Figure 7 are generated. In agreement with others who
have observed spectral changes upon annealing for PET (7–9),
characteristic bands for the trans isomer include 845 cm^{-1} (CH_2
rock), 973 cm^{-1} (C–O stretch), 1342 cm^{-1} (CH_2 wag), and 1472
cm^{-1} (CH_2 bend). Gauche bands at 897 cm^{-1} (CH_2 rock), 1370
cm^{-1} (CH_2 wag), and at 1472 cm^{-1} (CH_2 bend) appear in the
gauche spectrum.

Having obtained normalized gauche and trans spectra for
PET, least squares curve-fitting can be employed to determine

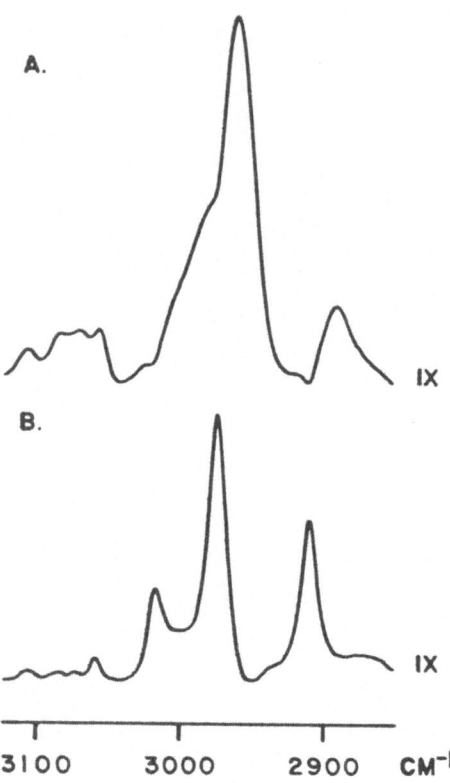

Figure 6.   PET: (A) Amorphous gauche C–H stretch; (B) crystalline and amorphous trans C–H stretch.

the percentage of each rotational isomer in films. Table 1 details
the results for a number of spectra taken of a single thin PET
film. The film was annealed at 104°C for up to 242 min. with
spectra taken at intervals as shown. Then the annealing temperature was raised to 150°C and additional annealing was done up
to 125 min. with spectra again taken at intervals. The film was
annealed further at 170°C for 1 hr. Tabulated in Table 1 is the
percentage of gauche and trans found in the films by curve-fitting.
If we assume the first spectrum is amorphous, we can follow the
amount of trans isomer in nucleation and/or crystalline regions

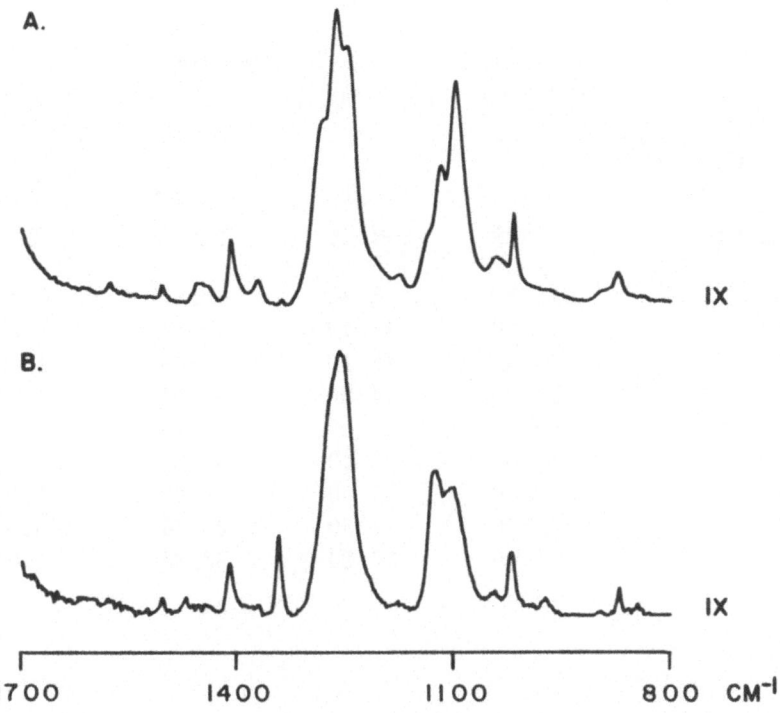

Figure 7. PET 1700-800 cm^{-1} region: (A) amorphous gauche; (B) crystalline and amorphous trans.

assuming the gauche/trans ratio in the amorphous phase remains constant and equal to that for the film nominally found for the film annealed for 92 min. at 104°C. The amount of crystalline trans increases to a maximum value of 46.5%.

We have, in conclusion, presented a systematic approach for investigating polymer chain conformations in the amorphous and semicrystalline materials. Fourier transform infrared spectroscopy has been utilized and combined with the techniques of factor analysis, absorbance ratioing, and least squares curve-fitting. The above approach was used to study amorphous and semi crystalline

Table 1

Ratio Method-Curve-Fit Results
PET Thin Films 1700-800 cm^{-1}

Cumulative Annealing Conditions	Total % Trans	Total % Gauche	% Trans (amor)	%Trans (xtal)
92 min 104°C	49.60	51.40	49.60	0.00
122	51.80	48.20	46.51	5.29
182	53.90	46.10	44.49	9.41
242	56.00	44.00	42.46	13.54
+5 min 150°C	65.70	34.30	33.10	32.60
35	68.60	31.40	30.30	38.30
65	70.00	30.00	28.95	41.05
125	72.50	27.50	26.54	45.96
+60 min 170°C	72.80	27.20	26.25	46.55

films of poly(ethylene terephthalate). Glycol trans and gauche conformational spectra were isolated and their volume fraction determined for a number of samples with different annealing histories. The success achieved with the application of the above techniques distinctly shows their value and characterization of amorphous and semicrystalline polymers.

REFERENCES

1. M.K. Antoon, L. D'Esposito, and J. L. Koenig, Applied Spectroscopy (submitted).

2. T.B. Hirschfeld, Analytical Chemistry, 48, 721 (1976).

3. J.L. Koenig, L. D'Esposito, and M.K. Antoon, Applied Spectroscopy, 31, No. 4, 292 (1977).

4. J. L. Koenig and D. Kormos (to be published).

5. M. K. Antoon, J. H. Koenig, and J. L. Koenig, Applied Spectroscopy, 31, No. 6, 518 (1977).

6. S. K. Bahl, D. D. Cornell, F. J. Boerio, and G. E. McGraw, Polymer Letters, 12, 13 (1974).

7. A. Miyake, Journal of Polymer Science, 38, 479 (1959).

8. L. D'Esposito and J. L. Koenig, Journal of Polymer Science, Polymer Physics Edition, 14, 1731 (1976).

9. I. M. Ward and M. A. Wilding, Polymer, 18, 327 (1977).

ACKNOWLEDGMENTS

The authors gratefully acknowledge the financial support received from the National Science Foundation under grant DMR77-01999-A01.

HIGH RESOLUTION CARBON-13 NMR STUDIES OF BULK POLYMERS

J. R. Lyerla

IBM Research Laboratory

San Jose, California 95193

I. INTRODUCTION

High resolution nuclear magnetic resonance (HR-NMR) spectroscopy is the apellative given to the measurement of NMR spectra under conditions that enable magnetically nonequivalent nuclei of the same spin species (e.g., protons or carbon-13) to be resolved as individual resonance lines. Such operating conditions are readily obtained for polymer melts and polymer solutions, and during the past 10-15 years, HR-NMR has evolved as a major analytical technique for the elucidation of structural features and motional characteristics of macromolecules in the liquid state. A list of the main parameters that are measured in NMR experiments is given in Table 1. The parameters have been classified as primary sources of either structural or dynamical data. Also listed are some of the specific types of information on polymeric systems that can be derived from these parameters.

For years, HR-NMR was synonymous with proton magnetic resonance in the liquid state. However, since the advent of commercial pulse Fourier transform NMR spectrometers in about 1971, there has been a rapid growth in the number of investigations in which the ^{13}C nucleus is employed as the probe of the molecular system. Typical of high resolution ^{13}C spectra of macromolecules in solution are those of the two stereoregular poly(methylmethacrylates) (PMMA) displayed in Figure 1. The spectrum of the isotactic polymer shows five resolved resonance lines, one for each carbon in the repeat unit. The spectrum of

the syndiotactic polymer also shows five dominant
resonance lines; however, the resonance position of each
carbon is shifted in frequency relative to its counterpart
in the isotactic spectrum. These differences in chemical
shift arise from tacticity effects and serve to illustrate
the sensitivity of carbon NMR to chain stereochemistry.
There are also present several less intense resonances
in the methyl and carboxyl regions of the spectrum of
the syndiotactic polymer. These arise from the presence
of a few meso (m) stereochemical relationships in the
predominantly racemic (r) chain. The three lines assigned
to the carboxyl carbon (whose chemical shift is sensitive
to pentad stereochemistry[1]) correspond to rrrr, rmrr and
rrrm stereo-sequences in the chain. The fact that there
are no resonances in the spectrum which arise from
sequences with adjacent meso dyads indicates the meso
dyads are introduced as isolated defects in the
polymerization.[2]

    Not only does the high degree of resolution in the
[13]C spectra of PMMA allow quantitative data on
stereochemistry to be determined, but it also makes it
possible to obtain directly data on sidechain and backbone
reorientational motion through the measurement of
spin-lattice, spin-spin and nuclear Overhauser relaxation
parameters for each carbon.[2]

    In contrast to the spectral definition of a few Hz
or less achievable in the liquid state NMR spectrum of
a polymer, the spectrum of a polymer solid (e.g., glassy
or crystalline materials) usually consists of a single
broad resonance line having a full-width at half-height
(FWHH) of tens of KHz. However, this result does not
imply that a solid spectrum is devoid of information.
To the contrary, study of the second moment and various
relaxation times of the resonance line as a function of
temperature and other parameters often yields important
data on the polymer such as degree of crystallinity,
chain axis orientation, and degree of molecular motion.
Indeed, so-called "wide-line" NMR has been one of the
primary methods of characterizing the bulk state of
polymeric systems.[3] Nonetheless, it might be assumed
that if a high-resolution NMR spectrum of a solid could
be obtained, new detail on the nature of the solid state
might be uncovered. Recent advances in magnetic resonance
have now made possible the partial realization of this

goal.[4] For example, recording of ^{13}C spectra having
linewidths in the range of 10-200Hz has been demonstrated
for polymers and organic solids. In this paper, our
purpose is three-fold: 1) to discuss the combination of
techniques required to obtain such ^{13}C spectra; 2) to
describe briefly the implementation of these techniques;
and 3) to illustrate the utility of the resulting spectra
in providing information on the composition of insoluble
polymers, chain orientation, motion in the solid state,
and other aspects of bulk polymers.

II. HIGH RESOLUTION ^{13}C-NMR IN SOLIDS

Preliminary to our discussion of ^{13}C-NMR in solids,
we mention several details of the chemical shielding and
the nuclear dipole interactions. Chemical shifts arise
because of the simultaneous interactions of a nucleus
with surrounding electrons and that of the electrons with
the applied magnetic field, H_0. The static field induces
electronic (orbital) circulation and polarization which
gives rise to a local magnetic field that opposes H_0 or
"shields" the nucleus from the full effect of H_0. Given
the nature of electron localization in chemical bonds,
it is readily apparent that rarely would the shielding
be spatially isotropic for a nucleus in a molecule.
Indeed, the shielding interaction is described by a
second-rank tensor, $\hat{\sigma}$ and Hamiltonian[5]

$$\mathcal{H}_{cs} = \hbar \sum_i \gamma_i \vec{I}_i \cdot \hat{\sigma} \cdot \vec{H}_0 . \tag{1}$$

In solution, rapid molecular tumbling results in
isotropic averaging of the shielding, reducing $\hat{\sigma}$ to the
scalar which defines the chemical shift observed in HR
studies. In the solid, the orientation dependence of
the shielding is not-averaged away.

The interaction between two nuclear dipoles is
described by a second rank tensor, $\hat{D}$. Because $\hat{D}$ is
traceless, the isotropic average is zero and thus the
interaction has no effect (to first-order) on liquid-state
spectra. In the solid, the magnitude of the interaction
is both orientation and distance dependent. The
Hamiltonian governing the interaction is given by[5]

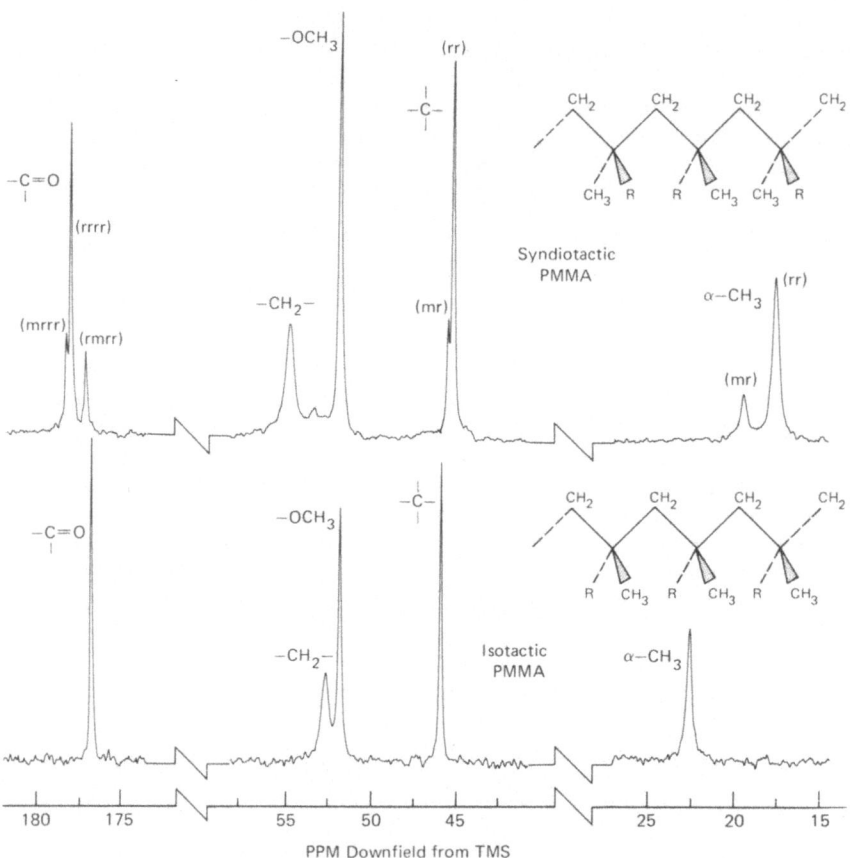

1. The 20 MHz proton-decoupled ^{13}C spectra of isotactic
and syndiotactic PMMA at 38°C in pyridine-d$_5$ (solvent
peaks eliminated). Each spectrum represents 7000
accumulated FID's. Symbols m and r refer to meso and
racemic stereochemical relationships between adjacent
repeat units. Structural representation of syndiotactic
and isotactic stereochemical triads are included above
the respective spectrum; the symbol R represents the
methyl ester side group. (Figure from reference 2.)

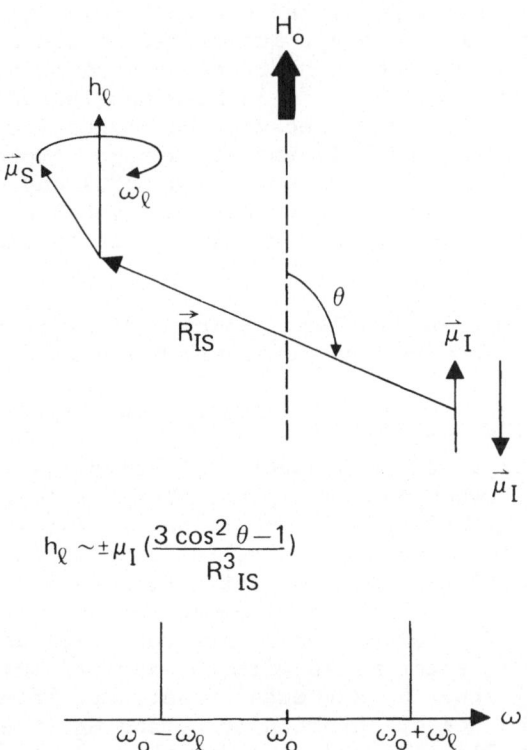

2. Schematic representation of an IS heteronuclear dipole
interaction. Symbols are defined in the text. The
frequency spectrum is given at the bottom.

$$\mathcal{H}_d = \hbar^2 \sum_{i<j} \gamma_i \gamma_j \vec{I}_i \cdot \hat{D}_{ij} \cdot \vec{I}_j \qquad (2)$$

where $\vec{I}_{i,j}$ are spin operators and $\gamma_{i,j}$ the respective magnetogyric ratios of the interacting nuclei.

We now turn to a discussion of the effects on the NMR spectra of solids produced by the orientation dependence of the dipolar and chemical shielding interactions. Since we are interested in the observation of carbon-13 resonance in hydrocarbon materials, we are primarily concerned with $^{13}C-^1H$ heteronuclear dipolar interactions. However, because the discussion is general (assuming spin-1/2 nuclei) for any system composed of a "rare" or low natural abundance spin (^{13}C) and an abundant spin (^1H), we employ the conventional symbolism S and I for the respective rare and abundant spin species.[6]

A. The Resolution Problem: Sources of Line-Broadening and Methods for Their Removal

1. <u>The I-S Dipolar Interaction</u>. The primary cause for the broad featureless resonance lines which are characteristic of the NMR spectra of solids is the dipolar interaction between nuclear spins. The dipole interaction between two unlike spins, I and S, having magnetic moments, μ_I and μ_S respectively, is depicted in Figure 2. The magnitude of the local magnetic field, h_ℓ, induced at nucleus S due to the magnet moment of nucleus I is a function of the internuclear separation, R_{IS}, and the angle, θ, this vector makes with the applied magnetic field, H_0. Whether h_ℓ augments or detracts from H_0 depends on the the orientation of μ_I and H_0 (i.e., the spin state of I); as a consequence, the resonance of the S nucleus is split into a doublet (except for $\theta=54.7°$ where $h_\ell=0$) whose members are shifted by $\pm\omega_\ell(=\mu_S h_\ell)$ from the resonance frequency, ω_0, obtained when the dipolar interaction is zero.

In a rigid polycrystalline system, a nucleus does not experience an isolated dipole interaction but many such interactions arising from neighboring nuclei having (in most instances) different R and θ relationships with the nucleus. The result is a Gaussian shaped resonance line centered at ω_0 and characterized by a second moment given by[7]

$$\langle \Delta\omega^2 \rangle_{SI} = 1/3\gamma_I^2\gamma_S^2\hbar^2 I(I+1) \cdot 1/N \sum_{j,k} \frac{(1-3\cos^2\theta_{jk})^2}{r_{jk}^6} . \quad (3)$$

Since the rigid lattice linewidth for a carbon bonded to a hydrogen is ca. 10-40 KHz, it is clear the width of the resonance line (except at very large applied fields) encompasses the entire range of carbon chemical shifts associated with diamagnetic organic structures. Of course, the linewidth is reduced if specific molecular motions are present to partially average[8] the dipolar interaction by imparting a time dependence to θ. In solution or the liquid state, rapid molecular tumbling results in isotropic averaging of $\langle 3\cos^2\theta-1 \rangle$ and hence a vanishing value of h_ℓ. The highly resolved spectra that are obtained reflect chemical shift and scalar (non-dipolar) spin-spin coupling interactions. To approach equivalent resolution in the solid state thus requires a method to force $h_\ell \to 0$. We now examine this situation for the carbon-proton case.

The truncated or secular dipolar Hamiltonian, $\mathcal{H}_d$, for two species I and S may be written as:[9]

$$\mathcal{H}_d = \mathcal{H}_{II}^0 + \mathcal{H}_{IS}^0 + \mathcal{H}_{SS}^0 \quad (4a)$$

where

$$\mathcal{H}_{II}^0 = \frac{\gamma_I^2\hbar^2}{2} \sum_{i}^{N_I} \sum_{i<j}^{N_I} r_{ij}^{-3}(3\cos^2\theta_{ij}-1) \cdot (\vec{I}_i \cdot \vec{I}_j - 3I_{iz}I_{jz}) \quad (4b)$$

$$\mathcal{H}_{IS}^0 = -\gamma_I\gamma_S\hbar^2 \sum_{i}^{N_I} \sum_{k}^{N_S} r_{ik}^{-3}(3\cos^2\theta_{ik}-1) \cdot I_{iz}I_{kz} \quad (4c)$$

$$\mathcal{H}_{SS}^0 = \frac{\gamma_S^2\hbar^2}{2} \sum_{k}^{N_S} \sum_{k<m}^{N_S} r_{km}^{-3}(3\cos^2\theta_{km}-1) \cdot (\vec{I}_k \cdot \vec{I}_m - 3I_{kz}I_{mz}) . \quad (4d)$$

The expressions for the various dipolar interactions have been written as a product of a spatial term and a spin term to facilitate discussion. As mentioned above, molecular tumbling isotropically averages the spatial term to zero in the liquid state. In a somewhat analogous

manner, i.e., by operating on the angular dependence of
the interaction, dipolar broadening can also be removed
in the solid state. If a solid sample is rotated at
frequency ω_r about an axis making an angle β with respect
to H_0 and angle α with respect to the internuclear vector
connecting interacting dipoles, it is straightforward to
decompose θ into α, β, and $\phi=\omega_r t$. This results in an
expression for the local dipolar field given by[10]

$$h_\ell = \pm\mu_j/R_{ij}^3[(3\cos^2\alpha-1)(3\cos^2\beta-1) + 3/2\sin2\alpha\sin2\beta\cos\omega_r t$$

$$+ 3/2\sin^2\alpha\sin^2\beta\cos^2\omega_r t] . \tag{5}$$

From Eq. 5 it is evident, h_ℓ consists of a static
contribution and a fluctuating contribution. If β is
chosen to be 54.7°, the so-called "magic-angle," the time
independent term vanishes and the spectrum consists of
a line at ω_0 (the Larmor frequency) having sidebands at
$\omega_0\pm n\omega_r$. The dipolar broadening is completely removed
from the central line and is manifest only in the
appearance of sidebands. For fast spinning rates, the
intensity of the sidebands becomes essentially zero and
the spectrum consists only of the narrowed line at ω_0.
It is important to note that $h_\ell\rightarrow 0$ irrespective of the
initial orientation of R and H_0, thus the dipolar
interactions in the molecular ensemble are removed.

We now inquire as to the rotational velocities
required for dipolar line-narrowing by magic-angle
spinning. Averaging of the dipolar Hamiltonian, which
is in essence the desired end result of the spinning
experiment, requires the rotation period of the sample
be short compared to the spin-spin relaxation time or
dephasing time, T_2, of the individual spins. Since T_2
is ca. the inverse of the linewidth, it is apparent that,
for effective removal of dipolar broadening, $\omega_r/2\pi$ should
be large compared to the dipolar linewidth. For
proton-proton and carbon-proton dipolar interactions,
this requires spinning speeds in the range 10-50 KHz.
While spinning rates as high as 15 KHz have been
achieved,[11] the upper limit is impractical as most
materials used to confine samples disintegrate due to
centrifugal forces. Thus, use of the MAS technique to
remove dipolar broadening, while successful in removing
homonuclear interactions for phosphorous and nuclei of

smaller magnetogyric ratio,[5] is of limited utility for proton homonuclear and heteronuclear interactions.

An alternative approach to removing the dipolar interaction which is particularly effective for the C-H case is to operate on the spin part of the Hamiltonian (Eq. 4) rather than the spatial term. If we focus on Figure 2, we see that one method of reducing the I-S dipolar term is to modulate the orientation of the I spins with respect to H_0. By forcing the spins to change states at a rate large compared to the frequency of the I-S interaction, h_ℓ is reduced to zero. This is accomplished by applying rf-irradiation at the I spin frequency of sufficient amplitude to "decouple" the interaction.[12] (In essence, this results in coherent averaging of the dipolar Hamiltonian in spin space as opposed to the incoherent averaging in molecular space due to Brownian motion in the liquid state.) Such double resonance techniques are standard in carbon magnetic resonance since high resolution spectra in solution are usually obtained by decoupling the ^{13}C-^1H scalar spin-spin interactions (J coupling). However, the rf power required is considerably greater to remove the 10-40 KHz dipolar coupling as compared to the 100-200Hz scalar coupling. For this reason, the removal of the I-S dipolar interaction by rf irradiation is referred to as "dipolar" decoupling[4] (DD).

Because of the 1.1% natural abundance of the ^{13}C isotope, the probability of having directly bonded two carbons carrying a magnetic moment is ca. 10^{-4}. Since the dipolar interaction attenuates rapidly (inverse third power) with distance between dipoles, and carbon has a small magnetogyric ratio relative to the proton, the homonuclear dipolar term, $\mathcal{H}^0_{SS}$, (Eq. 4d) is a negligible source of line broadening for carbons relative to the $\mathcal{H}^0_{IS}$ interaction. Thus, high power proton decoupling is particularly effective in reducing the linewidths of carbon-13 resonance lines in the solid state.

The third term in Eq. 4 is the homonuclear dipolar interaction for the I spins. In the absence of dipolar decoupling, exchange interactions which induce rapid "flip-flop" motion of the I spins can reduce the I-S broadening from its full rigid-lattice value. In this manner, the magnitude of I-I dipolar interaction can

affect the width of the S resonance. The details of this
spin exchange narrowing, which is analogous to classical
motional narrowing of resonance lines, is beyond the
scope of this Chapter and the interested reader is
referred to the literature.[13]

The effect of proton dipolar decoupling on the carbon
resonance of a solid is illustrated in Figure 3 for the
glassy state of poly-dioxydiphenylpropane-carbonate (the
polycarbonate of bisphenol-A). The spectrum in Figure 3a
(or lack thereof) was obtained under conditions of
proton-decoupling considered normal for measuring carbon
spectra in solution, while the spectrum in Figure 3b was
obtained using a decoupling field of ca. 32 KHz (7.5 gauss)
- ten times that used in Figure 3a. Removal of C-H dipolar
interactions allows resonances from the aliphatic carbons
to be distinguished from the aromatic carbons. However,
it is quite apparent there remains a residual broadening
of the resonances which inhibits greater resolution. We
now examine the nature of this remaining broadening. (It
should be noted that the apparent lack of even a broad
line spectrum in Figure 3a results from two factors: 1) the
low-abundance of the ^{13}C isotope and 2) the relatively
small number of free induction decays (FID) accumulated
for a case where intensity is dispersed over a several
KHz frequency range. Of course, the expected long
spin-lattice relaxation times in the solid make
repolarization of the carbon spins, and thereby addition
of free induction decays for signal/noise improvement,
an inefficient process. We will return to this point in
Section II.B.)

2. Chemical Shift Anisotropy. As discussed earlier
in this Section, the chemical shift is a second-rank
tensor which, because of rapid molecular tumbling, is
averaged to its isotropic value, i.e., 1/3 the trace of
the principal elements, in solution. In a polycrystalline
or amorphous solid, the orientation dependence of the
chemical shift is not averaged-out and as a consequence
can result in broadening of a resonance line. In
particular, unless the electronic shielding of a nucleus
is spherically symmetric, the chemical shift of the
nucleus will depend on the orientation of the principal
axes of the shielding tensor with respect to the magnetic
field direction. The NMR spectrum of a powder then
consists of a superposition of lines encompassing the

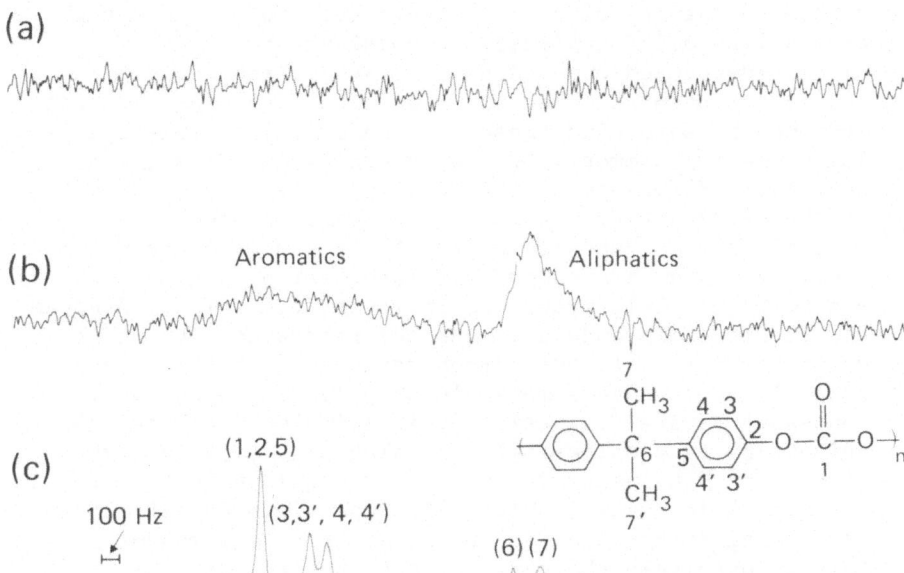

3. ^{13}C-NMR spectra of polycarbonate at 27°C: a) spectrum obtained under FT-NMR conditions normal for high resolution on polymer solutions; b) spectrum obtained under DD/CP conditions; c) spectrum obtained under DD/CP/MAS conditions. All spectra represent FT of 2000 FID accumulations. Resonance line assignments are given in the figure.

frequency range between the maximum and minimum values
of the shielding - in essence, a powder pattern. The
calculated lineshapes for an isotropic probability
distribution (powder) of tensor orientations for
uniaxially symmetric and non-axially symmetric tensors
are presented in Figure 4.[14] The principal values of
the shielding tensor, $\hat{\sigma}$, can be obtained directly from
the spectrum as indicated. In Figure 4 are the respective
carbon-13 spectra of solid benzene (at -40°C) and a highly
crystalline poly(ethylene), PE obtained under
dipolar-decoupled conditions. At -40°C, the rapid
reorientation of the benzene ring about the six-fold axis
reduces the shielding tensor to one of axial symmetry.
The non-axial symmetry of the shielding in PE is quite
apparent. Knowledge of the principal elements of $\hat{\sigma}$ and
of the effect of temperature (and thereby molecular
motion) on the shape and width of the powder pattern
provides a source of important information on a molecule[14]
- a point we shall return to in later discussion. Benzene
and PE (if chain ends are ignored) represent systems in
which there is only one carbon resonance; in the more
general case, where a molecule or polymer repeat unit
possesses several magnetically-inequivalent carbons, the
anisotropy in the chemical shielding will usually give
rise to an NMR spectrum composed of overlapping powder
patterns which make it virtually impossible to extract
structural information in any straightforward manner.
This is precisely the case for the spectrum of the
polycarbonate given in Figure 3b. To obtain a more highly
resolved spectrum requires the averaging of the various
carbon shielding tensors in the solid state.

The required averaging can be induced by utilizing
the magic-angle spinning technique introduced in
connection with the dipolar interaction. The chemical
shift of a nucleus i with principal shielding elements
σ_{11}, σ_{12}, σ_{13} and with direction cosines between the
principal axes of $\hat{\sigma}$ and $\vec{H}_0$, λ_{11}, λ_{12}, λ_{13} is given by[5]

$$\sigma_{izz} = \lambda_{11}^2 \sigma_{11} + \lambda_{12}^2 \sigma_{12} + \lambda_{13}^2 \sigma_{13} \ . \tag{6}$$

If the solid sample in which i is resident is rotated at
angular velocity ω_r about an axis inclined at an angle β
to $\vec{H}_0$ and at angles ϕ_{11}, ϕ_{12}, ϕ_{13} to the principal axes
of $\hat{\sigma}$, the direction cosines become

4. Calculated (N. Bloembergen and T. J. Rowland, Acta Metall. 1, 731 (1953)) absorption lineshapes and representative ^{13}C spectra for polycrystalline samples: a) an axially symmetric chemical shielding tensor ($\sigma_{||}$ and $\sigma_{\perp}$ denote shielding parallel and perpendicular to the symmetry axis) e.g., the spectrum of frozen benzene at -40°C; b) a general shielding tensor, e.g., the spectrum of polycrystalline polyethylene.

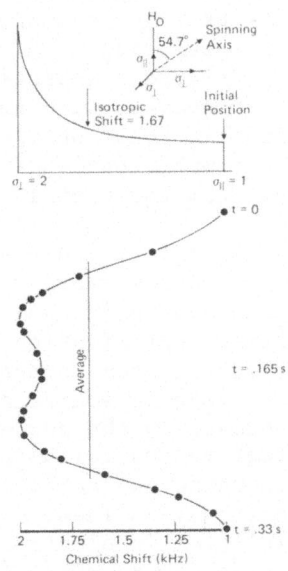

5. Plot of chemical shift vs. time during one rotor cycle at a spinning rate of 3 KHz for a carbon with an axially symmetric shift tensor having principal values $\sigma_{11}=\sigma_{22}=2$ KHz; $\sigma_{33}=1$ KHz. Data calculated using equations (6) and (7) for magic angle spinning (see diagram) and initial tensor orientation of σ_{33} parallel to H_0 and azimuthal angles of 0°/30° for σ_{22} and 90° for σ_{11}.

$$\sum_{k=1}^{3} \lambda_{ik} = \cos\beta\cos\phi_{ik} + \sin\beta\sin\phi_{ik}\cos(\omega_r t + \chi_{ik}) \qquad (7)$$

where χ_{ik} is the azimuthal angle of the kth principal axis at $t=0$. Substitution of (7) into (6) and taking the time-average yields

$$\bar{\sigma}_{izz} = 1/2\sin^2\beta(\sigma_{11}+\sigma_{12}+\sigma_{13}) + 1/2(3\cos^2\beta-1)\sum_{k=1}^{3}\sigma_{ik}\cos^2\phi_{ik} \,.$$
$$(8a)$$

If β is the magic angle (54.7°), (7) reduces to

$$\bar{\sigma}_{izz} = 1/3(\sigma_{11}+\sigma_{12}+\sigma_{13}) \qquad (8b)$$

which is the average value obtained in fluids. The result holds for all nuclei in the sample irrespective of the initial orientation of the principal axes of $\hat{\sigma}$ with respect to $\overline{H}_0$ and to the spinning axis. To illustrate the effect of spinning, Figure 5 depicts the chemical shift vs time plot for a nucleus with an axially symmetric shielding tensor during one full rotation of the sample about an axis making the magic angle with H_0. The average value of σ_{izz} obtained from the plot is $1/3(\sigma_{11}+\sigma_{12}+\sigma_{13})$.

As in the case of the dipolar interaction, the question again arises as to how rapidly the sample must be rotated to reduce the broadening due to chemical shift anisotropy. In the first approximation, the spinning frequency should be fast compared to the anisotropy expressed in frequency units or equivalently to the frequency range encompassed by the powder pattern. Typical shift anisotropy values for aliphatic carbons are 15-50 ppm while for aromatic and carbonyl carbons values of 120-200 ppm are observed.[15] These data reflect the larger directional variation in electron density (and hence shielding) associated with multiple bonds; this accounts for the broader resonance observed for the ring carbons as compared to the aliphatic carbons in Figure 3b. At an applied field of 1.4T, the anisotropy for aromatic and carbonyl carbons translates to ca. 1.8-3 KHz. Thus spinning speeds greater than 3 KHz would lead to removal of the broadening. In Figure 3c is the spectrum of polycarbonate obtained employing a combination of dipolar decoupling and magic-angle spinning at 4 KHz. The spectrum is sufficiently resolved that the majority of carbons in the repeat unit appear as individual resonance lines,

thus making it possible to derive detailed data on bulk
polymers. However, it is clear the resonance lines are
not as narrow as in a solution spectrum. We shall discuss
the sources of residual broadening and the limitations
to resolution in bulk polymers in Section IV.

B. The Sensitivity Problem

1. <u>General Considerations and Definitions</u>. We have
seen in the previous Section that central to the success
of the dipolar decoupling and MAS schemes in narrowing
carbon resonances was the fact that homonuclear dipolar
interactions between carbons could be ignored because of
the low isotopic abundance. Of course, the facility with
which carbon spectra can be observed depends strongly on
the application of digital methods of signal averaging
to overcome the "rare" spin sensitivity problem.
Acquisition of n free-induction decays (FID) results in
a S/N enhancement of $\sqrt{n}$. As is well-known, the rate at
which the FID following a $\pi/2$ rf pulse can be accumulated
is usually not limited by the digital hardware nor the
time for application of rf pulses, but by the rate at
which carbon magnetization (M_0^C) can be reestablished
along H_0 by spin-lattice relaxation processes. The
efficiency of T_1 processes depends on the spectral density
of motions at or near the Larmor frequency of the observed
nucleus (i.e., the MHz region). For polymers in solution
or the melt, carbon relaxation times are relatively short
- being a few hundred milliseconds at ambient temperatures
for carbons with attached protons and several seconds
for carbons without directly bound protons.[16] This occurs
because the molecular motion produces a high density of
the appropriate fluctuating local field needed to couple
the spin system and lattice and thus induce relaxation.
In contrast, motion at high frequencies in solids is
usually of small amplitude thereby resulting in a
substantially reduced spectral density in the Larmor
frequency region.[17] As such, the T_1 values for carbons
in solids may vary between a few seconds to minutes and
even hours, e.g., the carbon spin-lattice relaxation time
for crystalline polyethylene is ca. 700 seconds at ambient
temperature.[18] Clearly, obtaining a spectrum can be time
consuming if indeed not prohibitive in the case of
multi-line spectra. The situation is further aggravated
by the fact that linewidths in the solid spectrum may be

expected to be ca. 10-50Hz as compared to 1-5Hz widths
obtained for polymers in the melt or in solution. The
distribution of spectral intensity over a wider frequency
range manifests itself in requiring a greater number of
data acquisitions on the solid sample relative to the
liquid sample to obtain a spectrum having the same
"apparent" S/N (here we define S/N as peak height,
irrespective of width, divided by RMS noise).

As indicated in the above discussion, the carbon T_1's
in solids depend to some extent on the physical state of
the system. For example, in Figures 6a and 6c are shown
the ^{13}C spectra for a polycrystalline sample of the
material PHBA (polyhydroxy benzoic acid) and for a glassy
sample of polycarbonate, respectively. Both spectra were
obtained using conventional Fourier transform techniques, D D
and $\pi/2$ repetitive pulsing with delay times of 3.5 seconds.
Clearly, motion in the amorphous sample results in
reasonably short carbon relaxation times, T_1^C (although
the smaller amplitude for the lines assigned to carbons
without directly bound protons indicates full
repolarization does not take place for these carbons in
the repetition interval), while in the crystalline sample,
the lack of a spectrum suggests reduced molecular motion,
quite long relaxation times and little repolarization
between pulses. Fortunately, the situation for obtaining
spectra of crystalline systems is not as difficult as
this result would suggest. We now turn to some brief
comments on proton T_1 characteristics and on spin-lattice
relaxation times in the rotating frame.

2. <u>Proton T_1's in Polymers.</u> In solids, it is
generally the case that $T_1 \gg T_2$ - i.e., the coupling within
the spin system is much stronger than with the lattice;
hence, the spin system comes to internal equilibrium more
rapidly than it reaches equilibrium with the lattice.
As a consequence, if a spin system is perturbed from
equilibrium with the lattice, e.g., by a $\pi/2$ pulse, the
excess energy put into the spin system by the rf field
can remain for a time long compared to the time for the
spin system to establish internal equilibrium. However,
energy transfer to the lattice is an efficient process
near paramagnetic impurities, lattice imperfections or
near molecular segments which are in rapid motion (e.g.,
rotating methyl groups or amorphous regions in a
semi-crystalline polymer).[3] Even small concentrations

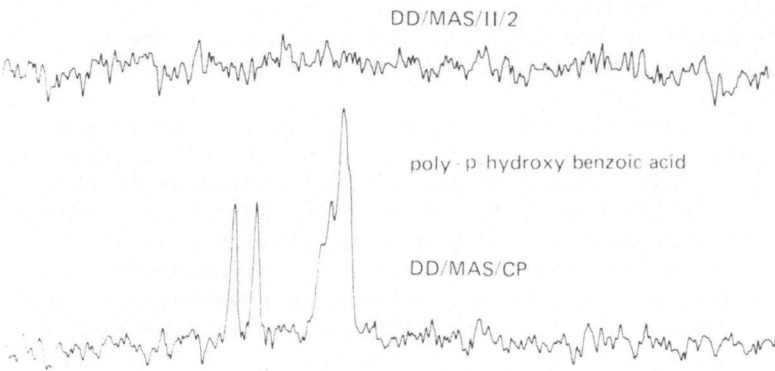

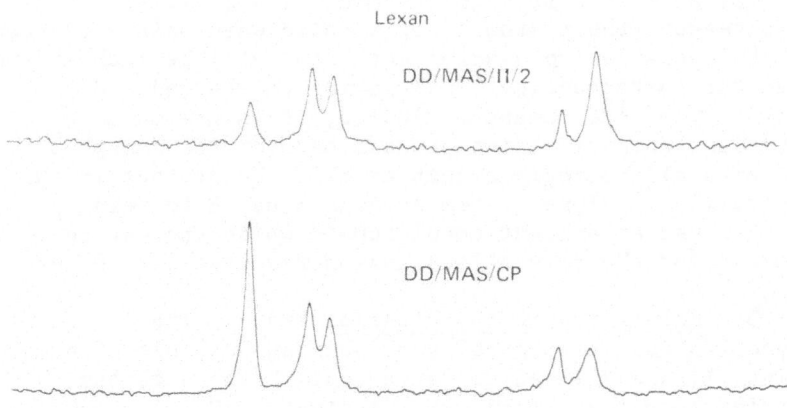

6. DD/MAS ^{13}C-NMR spectra of polycarbonate and PHBA. All spectra were taken with a 3.5s experimental repeat time and represent 1000 FID accumulations. In each case, one spectrum was recorded with 90° pulses and one with CP (CP time=1.5ms).

of these relaxation sites in a material can strongly
affect the overall rate of spin-lattice relaxation because
the excess spin energy can be transferred to these
preferred relaxation sites where, in turn, it is
transmitted to the lattice. In this manner, the entire
spin system comes to equilibrium with the lattice.

It is outside the scope of this discussion to explore
the details and complexities of spin diffusion. Put
simply, it is similar to any diffusion process in that
it involves transport driven by a concentration gradient.
In this instance, it is spin magnetization that is
transported by a spatial gradient in the magnetization.[19]
The magnetization does not diffuse by physical movement
of the nuclei or electrons having magnetic moments but
by mutual spin flips of <u>like</u> moments via the exchange
term of the dipolar Hamiltonian.[20] The coefficient of
spin diffusion is typically of the order 10^{-12} cm^2/sec
and depends on the distance between nearest like neighbors
and the probability of flip-flop transitions.[17] The
important aspect of the process for our present purposes
is the distance dependence, which enters the expression
for a diffusion-controlled[21] spin-lattice relaxation
process as the inverse third power, i.e., $1/T_1 \propto r^{-3}$. For
this reason, the proton T_1 of a solid when spin diffusion
is the operative relaxation mechanism will be much shorter
than the corresponding ^{13}C T_1 owing to the reduced
proximity of ^{13}C moments. Indeed, it is the more
efficient spin diffusion process between the amorphous
and crystalline regions that enables the protons in the
crystalline regions of semi-crystalline PE to relax in
ca. 1-5 sec at ambient conditions[22] while the carbon T_1
process for the crystalline region requires ca. 10^3 sec.

3. <u>Relaxation in the Rotating Frame</u>. The
magnetization (M_0^I) established when an ensemble of nuclear
spins, I, are allowed to reach equilibrium with the
lattice in a d.c. field, H_0, is given by Curie's law[23]

$$M_0^I = \frac{\gamma_I^2 \hbar^2 I(I+1)N_I H_0}{3kT} = \frac{CH_0}{kT} \tag{9}$$

where γ_I is the magnetogyric ratio, I the nuclear spin
quantum number, N_I the number of I spins per unit volume,
and C the Curie constant. Suppose spin-lattice
equilibrium has been reached by the I spins in the field

H_0. Further, suppose that suddenly (in a few microseconds) H_0 is reduced to a value $H_1 = 10^{-3} H_0$ and that quantized energy levels separated by $\omega_1 = \gamma_I H_1$ are established. Initially, the spin population distribution between the two energy states corresponds to that Boltzmann factor appropriate to H_0. However, by spin-lattice relaxation processes, the I spins will re-distribute themselves (by a net absorption of energy from the lattice) between the two spins states until the ratio of the occupation numbers is the Boltzmann factor appropriate to H_1 (i.e., equilibrium is obtained). Because in H_1, the two spin energy levels are separated by KHz frequency (rather than MHz frequency as in H_0), it is fluctuating local (dipolar) fields with components in this frequency regime that induce the necessary spin transitions for relaxation. Measurement of the spin-lattice relaxation time in H_1 thus allows information to be obtained on the density and amplitude of motions of the molecular system in the KHz range.[24]

The laboratory analogue of the proposed experiment is realized by a so-called "spin-lock" relaxation sequence. The sequence starts as a normal pulse-FT experiment on the I spins – i.e., the I spins, at equilibrium in H_0 (thereby having a net magnetization, M_0^I), are subjected to a resonant alternating rf field, H_{1I} perpendicular to H_0, of sufficient duration to rotate (in a classical vector representation) the net magnetization vector 90° ($\pi/2$ pulse). However, instead of now turning off H_{1I} and recording the FID, a 90° phase shift is introduced in H_{1I} which makes the magnetization vector and H_{1I} colinear in the rotating frame.[25] In this situation, (see Figure 7) the magnetization is said to be "spin-locked" along H_{1I}; the continued application of H_{1I} maintains the colinearity. The initial state of the spin system is exactly that described in the above paragraph, i.e., far from equilibrium in H_{1I} since $H_{1I} \ll H_0$. The subsequent relaxation process can be monitored by following the decay of magnetization M_1^I as a function of the time, τ, H_{1I} is applied after the phase shift. The decay of M_1^I toward equilibrium from its initial value of M_0^I is governed by the expression[26]

$$M_1^I(\tau) = M_0^I e^{-\tau/T_{1\rho}^I} \qquad (10)$$

where $T_{1\rho}^I$ is the spin-lattice relaxation time in the rotating frame[27] and depends on the spectral density at $\omega_1^I = \gamma_I H_{1I}$.

At this point it is instructive to examine the consequences of the spin-lock process in the context of the Boltzmann expression. At equilibrium in H_0, the observed magnetization $M_0^I \propto \exp(2\mu_I H_0/kT_S)$ where T_S is the temperature of the spin system[28] which is equal to that of the lattice T_L. After the spin-locking sequence, we have initially, $M_0^I \propto \exp(2\mu_I H_{1I}/kT_S)$. Since $H_{1I} = 10^{-3} H_0$, T_S must be $10^{-3} T_L$ for the condition to hold. In the language of spin thermodynamics, the result of the spin-lock sequence is to cool the I spins to near $0°K$. Of course, except at very low temperatures, there is in essence no change in the temperature of the sample because the heat capacity of the spin system is negligible relative to that of the lattice. The approach toward equilibrium in H_{1I} then corresponds to a heating of the spin system.

This lengthy prologue on proton relaxation times and spin-locking now puts us in a position to understand the solution to the ^{13}C sensitivity problem in solids proposed and demonstrated by Pines, Gibby, and Waugh.[14]

4. <u>Cross-Polarization</u>. Based on the above discussion, it is clear that often advantage could be obtained in the signal averaging process if the ^{13}C experiment could be recycled in the time required for proton repolarization ($\sim 5 \times T_1^H$) rather than for carbon repolarization ($\sim 5 \times T_1^C$). For this advantage to be realized requires the existence of mechanisms to establish carbon polarization via the protons despite the fact that in the static d.c. field, the large difference in resonance frequency of ^{13}C and ^1H results in essentially <u>isolated</u> spin systems. A solution to the problem of coupling the two spin systems was first recognized by Hahn.[29] This involves the application of two resonant alternating rf fields, H_{1I} and H_{1S}, to the respective I and S spins such that

$$\gamma_I H_{1I} = \gamma_S H_{1S} \ . \tag{11}$$

This condition, known as the Hartmann-Hahn condition,[29] corresponds to making the Zeeman splittings of the I-spins

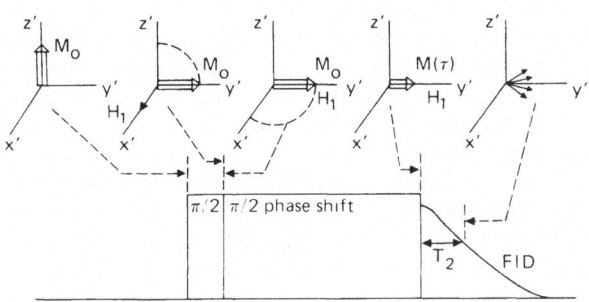

7. Pulse sequence used to measure the rotating frame spin-lattice relaxation time. (Figure from reference 3.)

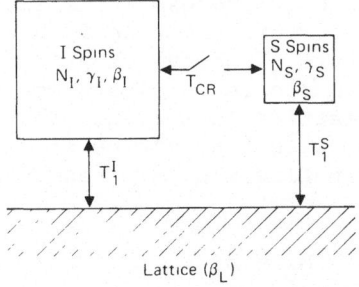

8. Schematic representation of an abundant I spin reservoir and a rare S spin reservoir, coupled to the lattice as indicated by their spin-lattice relaxation times, T_1^I and T_1^S. Coupling between the spin reservoirs is represented by T_{CR}, the cross-relaxation time. The coupling can be induced by a cross-polarization process in the rotating frame. (Figure similar to those of references 13, 14.)

quantized along H_{1I} in the I-spin rotating frame equal
to the Zeeman splitting of the S-spins quantized along
H_{1S} in the S-spin rotating frame. With Zeeman levels
matched, the spin systems can exchange energy via the
coupling produced by the I-S dipolar interaction.[30] A
classical description of the process is that the
precession of M^I about H_{1I} produces a sinusoidally
oscillating component of the dipolar field along the
direction of H_0. If the Hartmann-Hahn condition is
matched this frequency is correct to induce transitions
of the S-spins relative to H_{1S}. This method of polarizing
the S-spins is termed cross polarization (CP).

One of the variety of experimental methods[14] for
establishing contact of I and S spin systems and then
observing the S-spin polarization is represented by the
sequence:

1)   establish I-spin polarization in H_0
2)   spin-lock I-spins along H_{1I}
3)   contact I and S spins with a Hartmann-Hahn match
4)   turn-off H_{1S} and record S-spin FID
5)   repeat sequence.

In understanding the CP process it is instructive to
discuss the steps of this sequence from a simple
thermodynamic picture. Consider the I and S spin systems
to be two thermodynamic reservoirs which are in thermal
contact with the lattice (see Figure 8). Once the I-spins
are in equilibrium with the lattice (requiring a time
ca. 5 T_1^I), the magnetization is given by Eq. 9 which can be
re-written as

$$M_0^I = \beta_L C H_0 \; . \tag{12}$$

The relation $\beta = 1/kT$ defines an inverse spin temperature
which for the I spins at the end of step 1 is equal to
the inverse of the lattice temperature, β_L. Application
of the spin-locking sequence (step 2) effectively cools
(see previous discussion) the I spins to an initial
rotating frame temperature

$$\beta_0^I = (H_0/H_{1I})\beta_L \; . \tag{13}$$

The S-spins, which are assumed unpolarized (i.e., $T_1^S \gg T_1^I$),
are brought into contact with the I spins (step 3) and

the two spin systems come rapidly (in a time ca. T_{CR}) to
equilibrium at a common spin temperature, β_1. Because
the spin-lattice relaxation times are long, the two spin
reservoirs are essentially isolated from the lattice during
the cross-polarization process and thus the total energy
in the combined spin systems is conserved. Invoking
conservation, we find[14]

$$\beta_1 C_I H_{1I}^2 + \beta_1 C_S H_{1S}^2 = \beta_0^I C_I H_{1I}^2 \tag{14}$$

which allows us to determine β_1 as

$$\beta_1 = \beta_0^I (1+\varepsilon)^{-1} \tag{15}$$

where $\qquad\qquad \varepsilon = S(S+1)N_S / I(I+1)N_I$.

For the $^1H/^{13}C$ case, $\varepsilon \sim 0.01$ and thus $(1+\varepsilon)^{-1}$ can be
approximated by $(1-\varepsilon)$. The ^{13}C magnetization now
established in the rotating frame after one
cross-polarization contact is

$$M_1^S = \beta_1 C_S H_{1S} = \beta_I^0 (1-\varepsilon) C_S H_{1S} . \tag{16}$$

Re-writing (16) using (13) and invoking the Hartmann-Hahn
condition results in

$$M_1^S = \beta_L C_S H_0 (\gamma_I / \gamma_S)(1-\varepsilon) . \tag{17}$$

Thus the carbon polarization obtained in the CP process
is ca. 4× that established in H_0 (i.e., $M_0^S = \beta_L C_S H_0$).

After CP the contact between I- and S-spin systems
is broken by removal of the H_{1S} field and the FID of M_1^S
is observed (step 4). The S-spins then return to an
unpolarized state. The field, H_{1I}, remains on during
the recording of the FID to eliminate I-S dipolar
broadening. (The I-magnetization does not vanish in this
instance but decoupling is achieved since the axis of
magnetization is perpendicular to H_0.)

The transfer of polarization between I- and S-spin
systems is controlled by a cross-relaxation time constant,
T_{CR}. The rate (T_{CR}^{-1}) at which I- and S-spins reach thermal
equilibrium in the double-rotating frame depends, in

part, on the S-spin second moment due to I-S dipolar
interactions (see Eq. 3). The geometric dependence and
motional dependence of this interaction suggests that
1) different types of S-spins (e.g., carbons having
directly bound protons as opposed to non-proton bearing
carbons), 2) rigid groups as opposed to rapidly reorienting
groups (e.g., methyl groups), and 3) different orientations
with respect to H_0 of the same I-S spin pair (e.g.,
different portions of an anisotropy pattern) will reach
full equilibrium (see Eq. 17) at different rates. Thus
determination of T_{CR}^{-1} values by interrupting the CP step
after a time τ and observing the S spectrum as a function
of τ provides, in principle, a valuable source of
information on motion and orientation in solids.[14] We
will return to this aspect of the CP experiment in
Section IV.

From Eq. 15, it is apparent that the initial CP contact
results in only a small heating of the I-spins. Thus,
ideally, the remaining I-magnetization locked in the
rotating frame is

$$M_1^I = \beta_L C_I H_0 (1-\epsilon) = M_0^I (1-\epsilon) \ . \tag{18}$$

If $T_{1\rho}^I > T_{CR}$ (a necessary condition for cross-polarization
to occur, otherwise the I-magnetization will decay before
it can be transferred to the S-spins), it is possible to
repeat the contact and observe sequence (steps 3 and 4
above) before requiring I-spin repolarization (step 1).
If spin-lattice relaxation can be neglected, the S-spin
magnetization obtained from the nth contact will be[14]

$$M_1^S = (\gamma_I/\gamma_S)(1-\epsilon)^n M_0^I \ . \tag{19}$$

Co-addition of the S-spin's FID from each of the multiple
contact experiments yields, upon ultimate Fourier
transformation, a spectrum (from one spin-lock period)
having a total sensitivity enhancement even greater than
4×. For the interested reader detailed analyses of the
sensitivity considerations and of the CP process itself
are presented by Pines, et al.[14]

Although the development of a solution to the rare
spin sensitivity problem as described in this Section
proceeded with an investigation of cross-polarization on
the basis of a more rapid recycle time for experiments,

it is clear that the sensitivity enhancement obtained by
cross-polarization is also an important basis for
utilization of the technique. As an example, in Figures 6b
and 6d are the spectra of PHBA and polycarbonate using
the same recyle time as in Figures 6a and 6c but employing
cross-polarization in conjunction with MAS and DD. A
spectrum is now observed for PHBA and the intensities of
the polycarbonate spectrum are quantitative. The first
illustrations that combination of DD/CP/MAS techniques
could be used to obtain high-resolution ^{13}C spectra of
solids are due to Schaefer, Stejskal and Buchdahl[4,31,32]
who reported a series of results on glassy polymers. We
now turn to a brief discussion of experimental
implementation of the DD/CP/MAS techniques followed by
a discussion of their utility in studying macromolecules.

III. EXPERIMENTAL

In order for the reader to gain some insight into
the instrumentation required for DD/CP/MAS experiments,
a brief description of the spectrometer we have
employed[33,34] to obtain HR-NMR spectra in solids is given
in this Section. Particular emphasis is placed on the
MAS apparatus which is designed for variable temperature
operation.[35]

A. Dipolar Decoupling/Cross Polarization

The basic spectrometer is a Nicolet TT-14 pulse-FT
system operating at 1.4T with resonance frequencies of
15.087 MHz, 60.0 MHz and 56.4 MHz for ^{13}C, ^{1}H, and ^{19}F
respectively. The magnet is a Varian HA-60 electromagnet
with a 1 5/8" polegap. The spectrometer has been modified
for DD/CP by: 1) constructing a probe capable of
withstanding the rf power levels required; 2) adding
external gating circuitry (for spin-locking, etc.) and
rf amplifiers; and 3) rerouting some of the timing logic.
The probe (10mm i.d.) is a variation of the double-tuned
single coil design reported by Stoll, Vega, and Vaughan.[36]
The coil is tuned for ^{13}C resonance and either ^{1}H or ^{19}F
resonance. The probe has a "straight-through" design of
Dewar and an external deuterium (D$_2$O) lock system.
Broadband (0.25-105 MHz) 100W rf power amplifiers
(ENI-3100L) are employed to amplify the basic output of

the spectrometer's ^{13}C and ^1H channels or the low level
^{19}F output derived from a frequency synthesizer. With
this arrangement, it is possible to achieve a
Hartmann-Hahn CP match condition up to ca. 55 KHz for both
the ^{13}C-$\{^1H\}$ experiment and the ^{13}C-$\{^{19}F\}$ experiment.
If CP is not required, the proton decoupling network is
capable of sustaining a H_1 field of 22 gauss at 100W for
dipolar decoupling, while the ^{19}F network yields a
15 gauss H_1 field at 100W. A two-level decoupling scheme
was added to the spectrometer in order to apply a
continuous low level (ca. 1 gauss) H_{1H} field to the sample
and then gate-on the strong H_{1H} field for short periods.
When spectra are observed by $\pi/2$ pulses rather than CP,
the weak decoupling field saturates the protons thus
sustaining any nuclear Overhauser enhancement (NOE)[37]
present; application of the strong decoupling field during
observation of the FID provides for dipolar decoupling.
To remove artifacts in the CP spectrum which can arise
from the long rf pulses in the observe channel, data are
collected using a scheme[38] of alternating the proton spin
temperature.

B. Magic-Angle Spinning

Two main techniques have been used to produce the
high spinning rates required for MAS-one in which the
rotors are supported by gas bearings and the other in
which rotors are supported by an axle assembly. The
latter design, due to Lowe,[39] usually employs a
cylindrical rotor which spins on a horizontal
phosphor-bronze axle. The rotor is driven by compressed
gas directed from a jet onto peripheral vanes on the
rotor's exterior. The gas bearing design, pioneered by
Beams[11] and adapted to NMR by Andrew,[5] requires a
stator/rotor assembly with the rotors having a conical
underface and a cylindrical stack (see Figure 9).
Compressed gas enters the stator through an inlet tube
and is directed through inclined jets onto the rotor's
flutes.[5] The rotor is thus supported and driven by the
gas. The advantages of the Lowe design are the ease of
loading the sample cylinder into the rf coil of the probe,
ease of reaching the magic angle, and ease of adapting
the assembly to variable temperature operation.
Disadvantages of the design include rapid wear of the
rotor (a polymer) by the axle, sensitivity of spinning

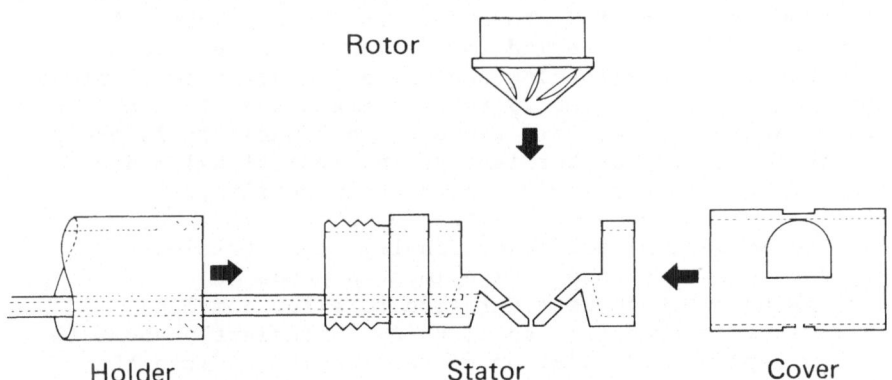

9. Schematic representation of the component parts of the spinner assembly and their relationship (see text for description). (Figure from reference 35.)

to rotor balance, and the small filling factor of the rf
coil volume by the sample. The advantages of the gas
bearing design are reduced wear, less sensitivity to
rotor balance, and a good filling factor as the cylinder
stack fits precisely into the coil (the cone of the rotor
is outside the coil) which is canted at the magic angle
rather than normal to the d.c. field.[40] Disadvantages
of this design include difficulty in sample loading (often
the coil itself must be removed), difficulty in the fine
adjustment of the magic angle condition, difficulty in
adapting the design to variable temperature operation.

The spinning assembly we have employed is a new
design[35] whose characteristics represent something of a
compromise of the two above-described techniques. The
general philosophy behind the design is to provide for
1) ease of sample loading and operation using conventional
designs of probes, and 2) for temperature variation. The
importance of a variable temperature capability is obvious
if features such as barriers to reorientational modes,
effects of phase transitions on chain mobility,
The nature of secondary relaxations, etc.
are to be determined. Additionally, the efficiency
of the double resonance experiment depends on
$T_{1\rho}^{I}$ being of sufficient duration to allow complete
polarization transfer and T_1^{I} being sufficiently short to
allow rapid repolarization of the I-spins. Since the
two relaxation times are functions of temperature, a
variable temperature spinning capability allows for an
optimally efficient[41] double resonance experiment to be
performed.

The spinner assembly is shown schematically in
Figure 9. The apparatus consists of four parts: holder,
stator, rotor and cover. The stator is threaded to the
holder and the gas line (in the case of ambient
temperature experiments) attached with a press fit. The
rotor is inserted into the stator cup and the cover sleeve
fitted over the stator. The general arrangement of stator
cup and rotor is similar to that of Andrew.[5] The gas
flow enters the conical "donut shaped" cavity behind the
stator cup (which is sealed to the outside by the sleeve),
is directed toward the rotor through three small
tangential holes (two of which are shown in the diagram)
drilled through the stator cup, and impinges on the rotor.
The arrangement of rotor flutes and stator holes is such

that a clockwise movement of the rotor is produced. The
exact shapes of these two components and the size and
angles of inclination of the holes for the driving gas
have to be carefully optimized as the apparatus makes a
very tight fit into the probe insert and thus exit gas
flow becomes important in stability considerations. The
spinning axis is at right angles to the holder and stator
assembly.

Spinner assemblies have been constructed from three
materials: Kel-F, Delrin, and machinable boron nitride
(BN).[83] For observation of hydrocarbon materials at
ambient and low temperature, the Kel-F assembly is used.
It displays suitable mechanical properties and does not
interfere with the carbon spectrum since the resonances
of the carbons in the Kel-F are >10 KHz in width due to
the unremoved C-F dipolar interactions. To observe
fluorocarbon materials at ambient and low temperature by
C-F dipolar decoupling/CP/MAS, the Delrin assembly is
used since unremoved C-H dipolar interactions broaden
the Delrin signal beyond detection. Finally, to obtain
high temperature spectra, the BN system, which is
transparent in ^{13}C-NMR, has been employed. Samples in
the form of powders are loaded into hollow rotors (volume
ca. 65μℓ) which are fitted with threaded caps to confine
the sample. In some cases, the polymer under study was
itself machined in the form of a rotor and used as the
analytical sample. The rotors have a cone diameter of
8mm and experience centripetal acceleration on the
perimeter of ca. 0.4 million G when rotating at 5 KHz.[42]

Since variable temperature operation down to at least
liquid nitrogen temperature is desirable, we have utilized
helium as the propellant gas. At ambient temperature
and 15 p.s.i. pressure, spinning rates >5 KHz are readily
achieved. (The maximum theoretical rate for a 8mm
diameter rotor with He as the propellant is ca. 38 KHz at
0°C.)[43] These rates are obtained on rotors without
flutes. As fluting a rotor is a complicated machining
operation, the ability to utilize smooth rotors is a
desirable feature. Temperatures below ambient are
achieved by mixing He at ambient temperature with He
precooled by flow through a heat exchanger immersed in
liquid nitrogen. The mixture of gases is transferred by
a stainless steel Dewar (over which the spinning assembly
fits) directly into the stator cup resulting in spinning

and cooling. Temperatures above ambient are obtained by
pre-heating the He at entry into the transfer Dewar. To
the present, we have operated over a temperature range
of -195°C to 100°C.

For an electromagnet, there are two alternative ways
in which the "magic-angle" arrangement can be achieved
by continuous adjustment using the spinner design. The
correct angle can be obtained by inserting the holder
vertically into the probe (with the rotor spinning around
a horizontal axis) and rotating it until the correct
angle is reached, or by inserting the holder horizontally
into the probe (also mounted horizontally) with the rotor
spinning around a vertical axis and then rotating the
holder until the correct angle is reached. Although the
first arrangement in reaching the angle corresponds to
a conventional rf-coil/H_0-field geometry, the variable
temperature hardware is more easily arranged with the
second geometry and, as such, is used on our spectrometer.
In both arrangements, the exact angle chosen is reached
by continuous adjustment of the rotation angle which is
controlled by a goniometer head attached to the top or
front of the probe body. This gives considerable fine
control over the exact angle chosen in an experiment.

In addition to the spinner assembly detailed here,
several other variations of Andrew-type spinners have
recently appeared in the literature.[10,11,44] Also
commercial instrumentation to carry out DD/CP/MAS
experiments is now entering the marketplace.[45]

IV.   APPLICATIONS OF ^{13}C NMR TO THE SOLID STATE OF
                    MACROMOLECULES

A.   Structure and Composition

In Figure 10 is shown the carbon-13 spectrum obtained
from a polycrystalline sample of the homopolymer of
p-hydroxybenzoic acid (PHBA).[46] The spectrum is resolved
into three distinct resonances of 1:1:5 relative
intensities with the resonance of five-fold intensity
having partially resolved components. The various carbons
of the repeat unit are readily assigned to specific
resonances on the basis of model compound studies and
cross-relaxation studies (see below). Clearly, the

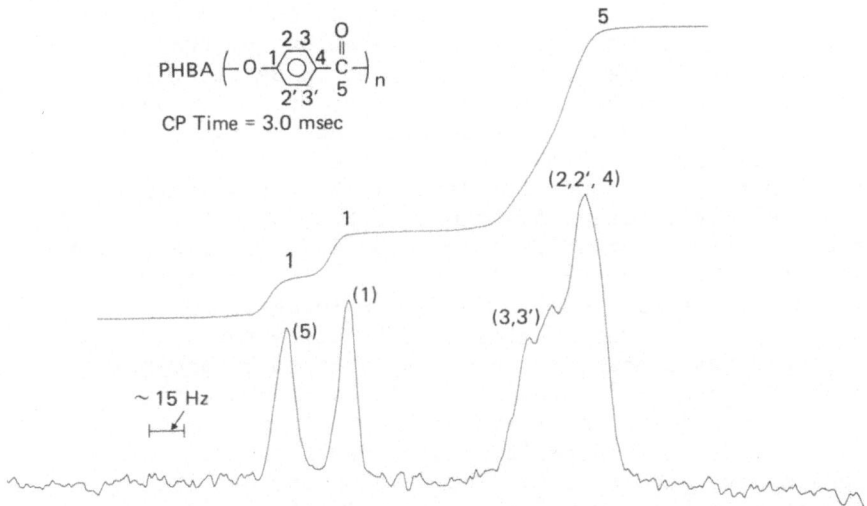

10. DD/CP/MAS ^{13}C-NMR spectrum of PHBA at 27°C. Spectrum obtained from 2K FT of 1500 FID accumulations with a CP time of 2.0ms and an experiment repetition time of 3.5s. Line assignments are given in the figure and the spectral integration is displayed.

spectrum represents a quantitative "fingerprint" of the
polymer in the bulk state and makes possible the study
of molecular dynamics at individual sites in the molecule.
The PHBA polymer is insoluble in all common solvents and
therefore not subject to solution characterization. This
capability to analyze insoluble and cross-linked polymers
promises to be one of the more useful aspects of the
solid state NMR techniques.

Another important facet of the PHBA spectrum is the
appearance of separate resonances for the C-3 and C-3'
carbons. The assignment of the resonances is definitive
based on the sequence of cross-relaxation spectra
displayed in Figure 11. As explained in Section II.B.4,
the rate of cross polarization depends primarily on the
magnitude of the static C-H dipolar interaction
experienced by a carbon. Thus, in the first
approximation, carbons with directly-bonded protons
polarize more rapidly than those without direct
interactions.[47] At CP times of 0.02 and 0.05 msec, only
resonances from the two sets of ortho ring carbons are
present in the spectrum. The 1:1:2 intensity pattern
and the value of the chemical shifts allow the two
downfield resonances in the two spectra to be accorded
to the C-3 and C-3' carbons. At longer contact times,
the region between the lines of the 2,2' and 3,3' carbons
begins to develop intensity and at ca. the same rate as
the line assigned to C-1. The development of line
asymmetry in the 120-135 ppm region is clearly due to C-4
and demonstrates that this carbon resonates at 125.4 ppm,
downfield of carbons 2 and 2'. The spectra at longer CP
times are also in agreement with the proposed assignment
of the carboxyl carbon in that the cross-polarization
rate of the C-5 resonance is the slowest as is expected
on the basis of a larger distance to protons.

The non-equivalency of the C-3,3' ring carbons is
direct evidence that the motion of the ring is confined
to small amplitude oscillations about an asymmetric
equilibrium position. The spectrum has been examined in
the temperature range -155°C to +80°C and the separation
of the C-3,C-3' carbons remains constant at 51Hz. The
80°C spectrum allows a minimum barrier to rotational
averaging of >17 Kcal/mol to be calculated. We are
currently attempting to obtain spectra at higher
temperatures using the BN spinning assembly in order to

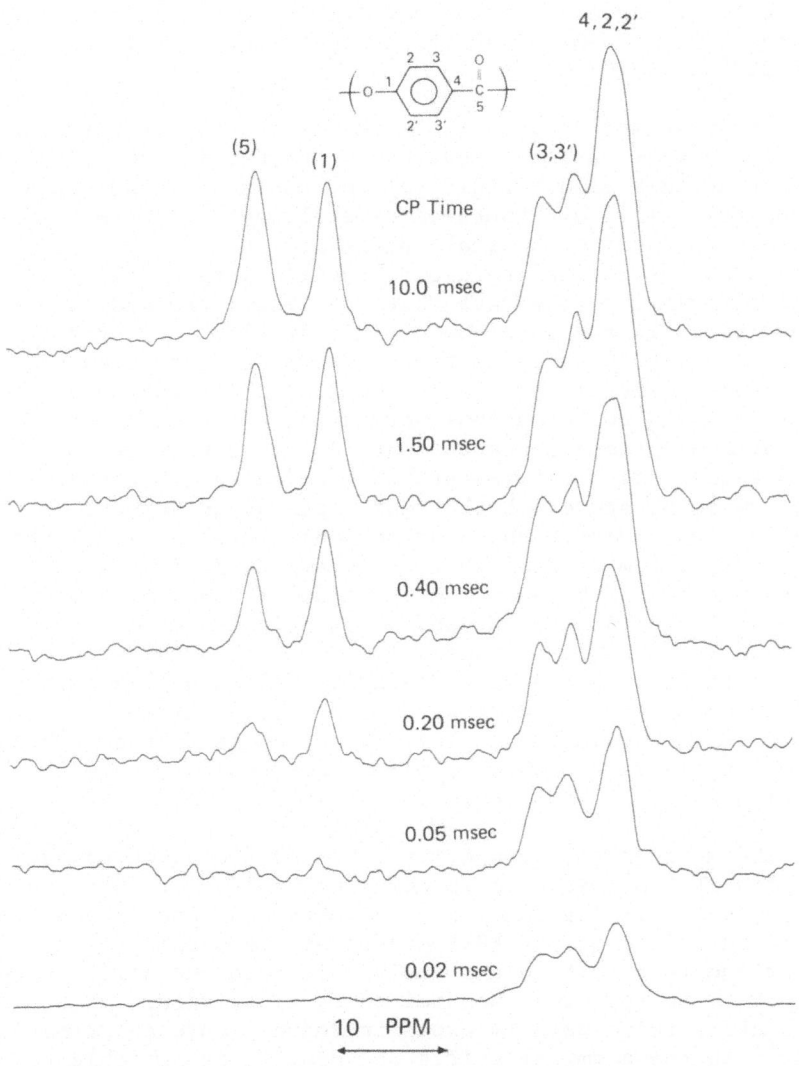

11. Magic–angle CP ^{13}C–NMR spectrum of PHBA at 27°C as
a function of CP contact time (in msec). All spectra
obtained from 2K FT of 1000 FID accumulations at a 3.5 sec
experiment repetition time. (Figure from reference 33.)

observe the coalesence of the two lines and determine
activation parameters. However, dielectric data[46] on
the polymer suggests motion sufficient to cause averaging
of the C-3,3' chemical shifts may not occur until
325-350°C.

The proposed structure of PHBA is a double helix with
the two chains having a reversed head-to-tail order,[46]
resulting in a highly rigid polymer chain. The NMR data
supports a rigid structure in several ways: 1) the
non-equivalent ^{13}C chemical shifts of the 3,3' carbons
indicates the limited torsional oscillations of the
aromatic ring; 2) the fact that the linewidths and shift
separations in the spectrum of PHBA at -155°C are the
same as at +80°C suggests that temperature reduction does
not produce any dispersion in chemical shift due to
non-averaging conformations nor any significant change
in lattice motions since the marginal decoupling field
(6.5 gauss) still removes proton broadening effects at
the low temperature - both these conclusions support a
highly rigid polymer chain at ambient temperature; 3) the
proton spectrum of PHBA which is characterized by a
linewidth of ca. 6 gauss and T_1 value of 3-4 sec at 25°C is
also unchanged at -155°C; and 4) as pointed out in
Section I.B.1 it is not possible to obtain a spectrum of
PHBA with DD, MAS and $\pi/2$ repetitive pulses on the carbon
spin system using as much as 20 sec recycle times and
8,000 FID accumulations. This result suggests the carbons
have long T_1 values which implies only small spectral
density in the 15 MHz region.

The magic-angle CP spectra obtained from 25mg powder
samples of two insoluble methoxy derivatives of PHBA are
shown in Figure 12a and 12b. The perturbations in the
electron shieldings of PHBA caused by the methoxy
substituents result in extremely well-resolved solid-state
spectra. By applying chemical shift substituent
parameters for a methoxy group introduced onto an aromatic
ring[48] to the chemical shifts of PHBA, it is possible to
tentatively assign the resonance lines in Figure 12 to
specific carbons of the repeat units. The observed line
intensities indicate that at the chosen value of
cross-polarization time, the spectra are essentially
quantitative. The apparent non-equivalency of the methoxy
carbons and 3,3' ring carbons in the dimethoxy polymer
indicates restricted reorientation of the aromatic ring

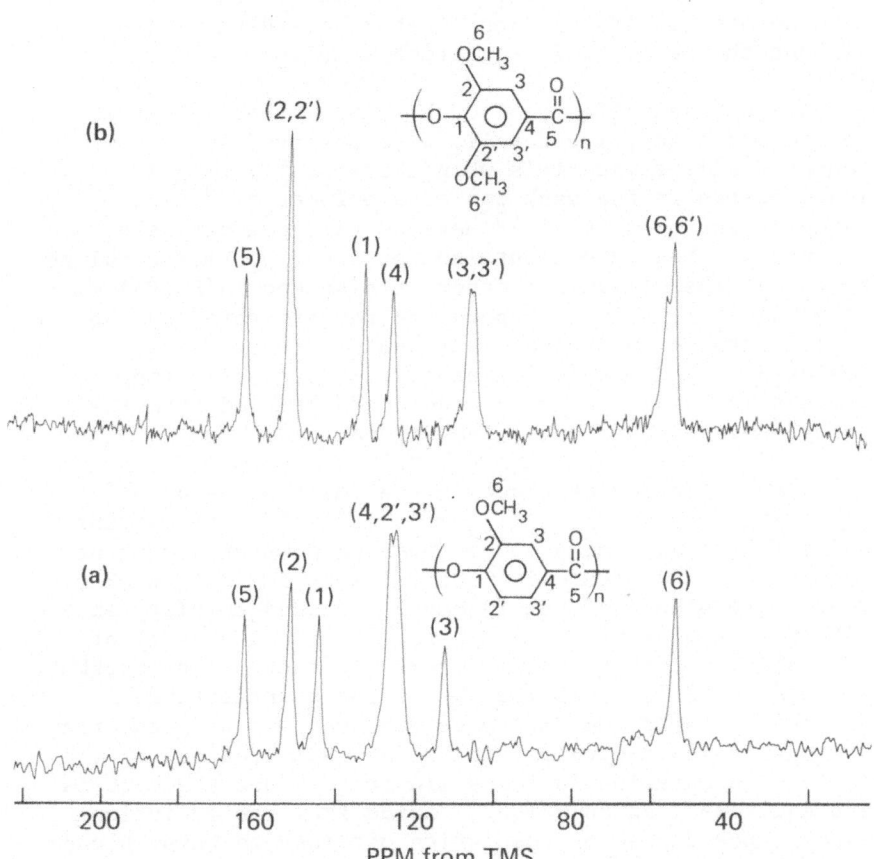

12. a) Magic-angle CP ^{13}C-NMR spectrum of a mono-methoxy derivative of PHBA at 27°C. Spectrum obtained from 2K FT of 21,900 FID accumulations with a CP contact time of 2.5 msec and experiment repetition time of 3.0 sec. b) Magic-angle CP ^{13}C-NMR spectrum of a dimethoxy derivative of PHBA at 27°C. Spectrum obtained from 4K FT of 19,000 FID accumulations with other conditions as in a). (Figure from reference 33.)

and thereby the polymer backbone. Because the monomethoxy
polymer would be expected to have a repeat unit composed
entirely of non-equivalent carbons, it is not possible
to make any qualitative conclusions regarding motion
without the benefit of relaxation studies.

The non-equivalency of ortho ring carbons observed
in the aromatic polyesters is also observed in
non-crystalline materials when the aromatic ring is
incorporated in the backbone of a polymer.[4,49] For
example, Garroway et al.[49] have carried out variable
temperature MAS experiments on the epoxy polymer obtained
by curing the diglycidyl ether of bisphenol A(DGEBA) with
piperidine. The 100° temperature range covered by the
experiments is sufficient that restricted rotation,
coalesence, and motional averaging of the ortho ring
carbons in DGEBA that are three bonds removed from the
methylene are observed. (See Figure 13.)

The resolution attained in the solid state by
employing DD/CP/MAS techniques is sufficient to warrant
using NMR to determine composition of polymer blends and
copolymers. In Figure 14, the ^{13}C spectrum of one of the
compatible blends of poly(phenylene oxide)/poly(styrene)
(PPO/PS) is presented. Comparison of the intensity of
the methyl carbon resonance of PPO to that of the backbone
carbons of PS is an obvious basis for quantitative
analysis. As in the PHBA spectrum, one set of ortho ring
carbons is non-equivalent in PPO - again, reflecting the
restricted rotation in the solid state since the carbons
are equivalent in solution.[4] Aside from the analytical
aspect, the degree of resolution obtained on these blends
may allow insight to be gained into the nature of the
interactions responsible for compatibility of the two
homopolymers, e.g., from variable temperature
heteronuclear cross-relaxation and spin-lattice relaxation
studies as a function of blend composition. An example
of copolymer analysis is provided by the spectra of two
insoluble p-hydroxybenzoic acid/biphenylene terephthalate
copolymers given in Figure 15. The monomer feed ratios
for the compounds were 2/1 and 1/2 molar ratios
respectively. Although the spectra show considerable
overlap from the many aromatic resonances, the lines
corresponding to the carboxyl carbons and the aromatic
carbons bonded to oxygen are resolved. The resonance of
the carbons bonded to oxygen in the HBA units differs in

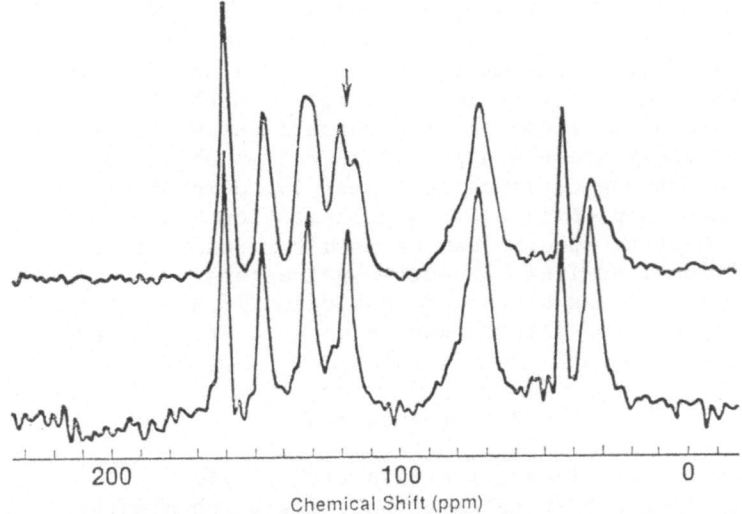

13. ^{13}C DD/CP/MAS spectra of piperidine cured epoxy at
−36°(top) and 58°C(bottom). The splitting in the aromatic
ring carbons (−36°C) denoted by the arrow is 5.5 ppm.
(Figure from reference 49.)

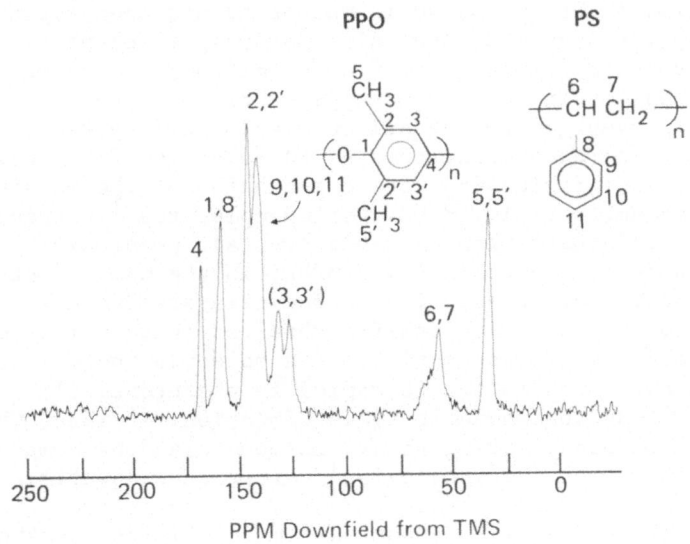

14. The DD/CP/MAS ^{13}C-NMR spectrum of a blend of
polystyrene/polyphenylene oxide at 27°C. Assignments
are indicated in the figure.

chemical shift from the corresponding carbons in the BPT
units. For the 2/1 copolymer the two resonance lines
would be of equal intensity since the BPT unit has two
equivalent carbons of this type, while in the 1/2
copolymer the ratio of the resonance lines would be 1/4.
Clearly, these are the approximate ratios observed for
these two resonance lines in Figure 15; demonstrating
that, under appropriate cross-polarization conditions,
the solid-state spectra can be used for quantitative
compositional analysis on such copolyesters.
Unfortunately, the degree of resolution is not sufficient
in the spectra to allow sequence data to be obtained.

B.   Orientation

Aromatic copolyesters of HBA with ethylene
terephthalate are examples of polymers which display
liquid crystalline order - in this case, thermotropic
liquid crystalline[50] behavior is exhibited (for certain
compositions) in the melt. Mesomorphic order in polymers
is of current interest because of the technological
implications with respect to ultrahigh strength/modulus
materials.[51] Proton magnetic resonance has been employed
to study the nature of low molecular weight mesophases
for many years.[52] The use of DD/CP methods to observe
^{13}C resonances in mesophases appears to be capable of
providing even greater detail on mesomorphic systems,
since dipolar decoupling removes non-averaged C-H dipolar
interactions arising from the orientation of the molecules
in the magnetic field.[53-55] For example, the ^{13}C spectra
of the model thermotropic liquid crystal, p-methoxy
benzylidene-p'-n-butylaniline, MBBA, in the nematic state
as a function of cross-polarization time are shown in
Figure 16. The chemical shifts observed between isotropic
and nematic states are typical of a molecule whose long
axis (about which motion is rapid) is preferentially
aligned along the magnetic field direction.[53] Thus, the
aromatic carbons, having their maximum shielding component
perpendicular to the long axis,[15] experience downfield
shifts in the nematic state relative to the isotropic
state. The side-chain carbons are more shielded in the
long-axis direction and thus experience upfield shifts;
however, the shifts are much less than for the aromatic
carbons owing to the smaller shielding anisotropies for
the aliphatics and to a lesser degree of ordering in the

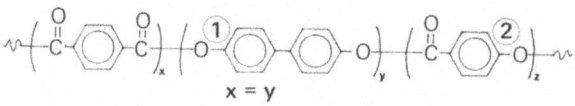

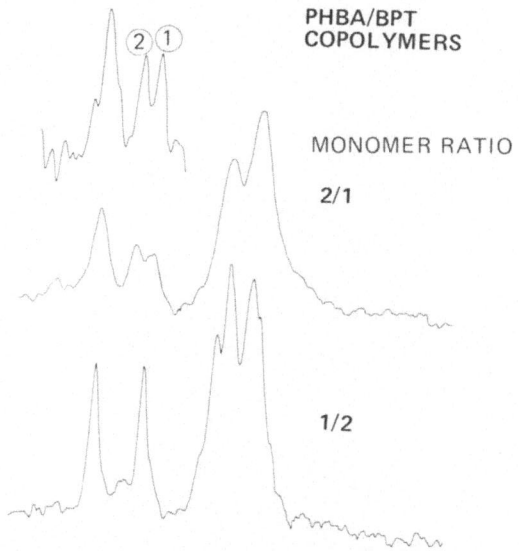

PHBA/BPT
COPOLYMERS

MONOMER RATIO

2/1

1/2

15. Magic-angle CP spectra for 2/1 and 1/2 HBA/BPT copolymers at 27°C. Spectra were obtained with CP times of 3ms and recycle times of 2.5s. The spectra are the result of 8000 accumulations. The inset spectrum for the 2/1 copolymer is one obtained with better setting of the magic-angle spinning condition. (Figure from reference 33.)

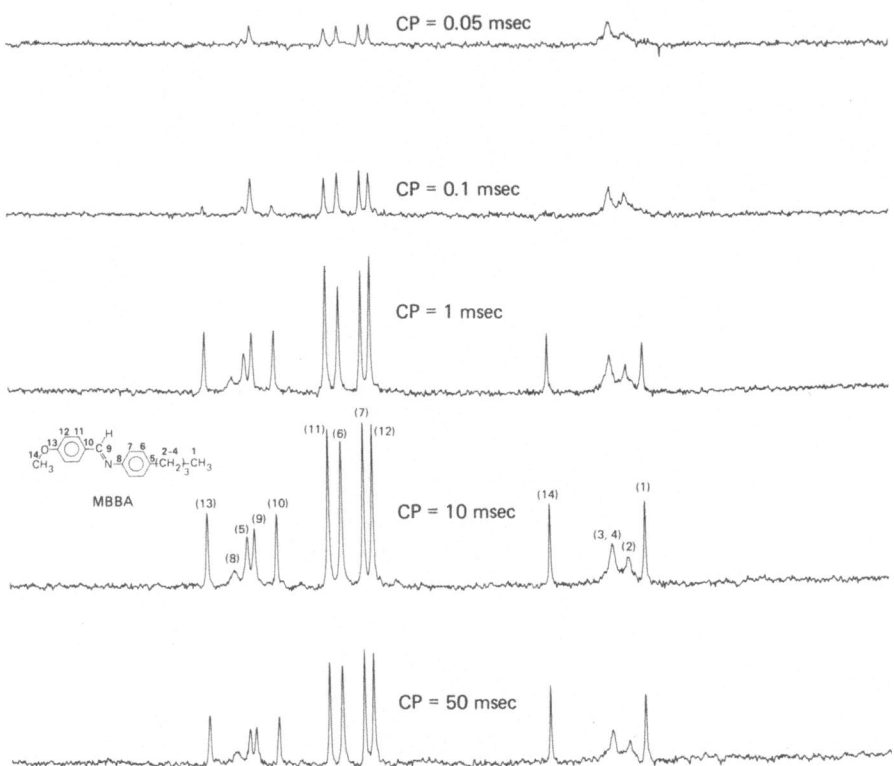

16. ^{13}C-NMR spectra of MBBA in the nematic phase at 22°C as a function of cross-polarization time. Resonance line assignments are indicated in the spectrum obtained with CP=10 msec. Each spectrum represents the FT of 400 FID accumulations. Results are reported on a normalized scale.

aliphatic tail. The dynamics of polarization build-up
indicate that the methoxy and methyl carbons have quite
reduced residual C-H dipolar interactions as compared to
the carbons of the aromatic core. This is, no doubt,
the result of chain flexibility and internal rotation
about C_3-axes. Orientation data on the degree of
alignment of the long axis with H_0 is also available in
the spectra. In the case of smectic-A mesophases, the
DD/CP spectra are strongly dependent on orientation of
the "director axis" and the field. A parallel orientation
can be established by cooling in the magnetic field
through a nematic into a smectic phase. Once the smectic
phase is formed under the influence of H_0, the high
viscosity of the smectic phase, unlike the nematic,
prevents realignment of the long-axis along H_0 when the
sample is rotated thus producing a spectral orientation
dependence characteristic of a uniaxial crystal.[53-55]
The extention of these techniques to polymeric liquid
crystal compounds seems straightforward and offers the
potential to learn about the degree of ordering in such
systems, composition of coexisting phases, and molecular
motion.

Obviously, the spectra of the liquid crystalline
systems are obtained without MAS since rapid spinning
would destroy the orientation data. Indeed, as was
alluded to in Section II.A.2, the improvement in resolution
gained by magic-angle spinning is at the expense of the
information contained in the chemical shift anisotropy.
The significance of the directional dependence of the
chemical shift in elucidating polymer structure is
evidenced by the spectra of PE obtained under
dipolar-decoupled conditions (Figure 17). The principal
elements of the shielding tensor (51.4, 38.9, 12.9 ppm
from TMS) have been assigned with respect to the geometry
of an all trans polymer chain.[56,57] Based on these
assignments the carbons of a PE chain in which the chain
axis is aligned with H_0 will resonate at 12.9 ppm. As
demonstrated by VanderHart,[58] this provides a method of
monitoring the chain orientation induced in a PE sample
by processes such as drawing. The ^{13}C spectrum of an
oriented PE sample in which the sample was configured in
the spectrometer such that the draw direction is parallel
to H_0 is given in Figure 18a. The bulk of intensity from
the sample (which was prepared by extrusion at 2400 atm
and 130°C through a conical dye to produce a draw ratio

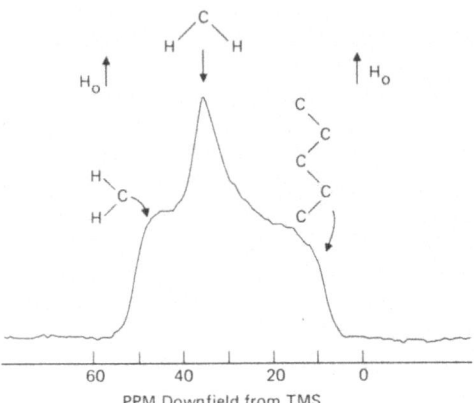

17. The DD/CP ^{13}C spectrum of polycrystalline PE. The three elements of the chemical shielding tensor and their assignment in an all-trans chain are indicated.

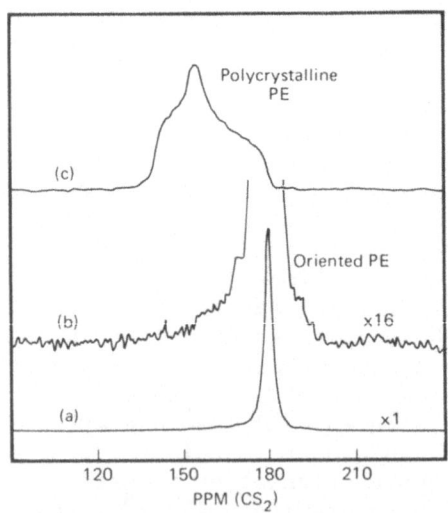

18. The DD/CP ^{13}C-NMR spectra of: a) fiber oriented PE, draw direction fixed parallel to H_0; b) as in a but ×16 amplification; c) a polycrystalline sample of PE. (Figure from reference 59.)

ca. 12:1)[58] is centered at 12.9 ppm, consistent with a high
degree of chain orientation in the draw direction.
Examination of the spectrum under higher amplification
(Figure 18b) shows intensity is spread over the entire
38 ppm chemical shift range encompassed by polycrystalline
PE, indicative of the presence of misaligned chains in
the oriented sample.

Spectra of the oriented PE sample as a function of
temperature[58] are given in Figure 19. Several features
are of interest: 1) As the temperature is raised, there
is a significant decrease in the width of the downfield
shoulder on the central resonance line. This result
indicates that the misaligned chains show greater motional
averaging than the oriented chains, as might be expected
on the basis of poorer packing in the regions of
misaligned chains. 2) At high amplification, satellites
are present in the spectra which arise from $^{13}C-^{13}C$
dipolar couplings. Because of the sample's high degree
of orientation, the resonances occur at discrete positions
such that the satellite splitting is given by

$$\Delta = 1.5\gamma_C^2 \hbar^2 (1-3\cos^2\theta_{ij})/R_{ij}^3 . \qquad (20)$$

Calculation of the satellite splittings[58] based on an
all trans PE chain indicates the three strongest
interactions (marked by the dashed lines in Figure 19)
arise from pairs of carbons on the same chain which are
one, two, and three bonds separated. Only for the
directly bonded carbons is the pair of satellites
observed, the downfield members of the two and three bond
interactions being obscured by the resonances from
misaligned chains. The breadth of the satellite
resonances arises from the presence of misaligned chains
and the narrowing that occurs at 102°C is most easily
interpreted as resulting from rotational averaging of
θ_{ij} about the chain axis for those chains which are
slightly misaligned relative to H_0.

The restricted reorientational freedom in the
crystalline regions of PE results in spin-lattice
relaxation times of ca. 10^3 sec.[59] While CH_2 carbons of
PE chains in amorphous material have T_1's ranging from
~0.3s to 3s.[60] Advantage can be taken of this
differential in relaxation rates to study only the
non-crystalline regions of PE. VanderHart[59] has shown

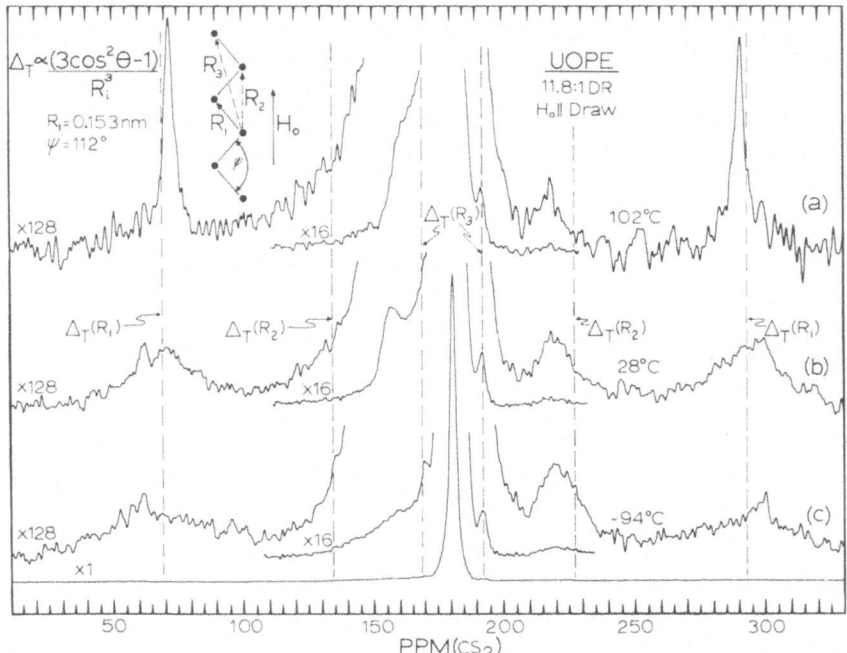

19. ^{13}C cross-polarization spectra of ultraoriented
linear polyethylene with an 11.8:1 draw ratio. The draw
direction is parallel to the magnetic field. The three
pairs of ^{13}C-^{13}C dipolar satellite positions based on
R_1=0.153nm and ψ=112° are given by the dashed lines.
Temperatures and spectral amplification factors are also
given. (Figure from reference 58.)

that by performing a 180-t-90 pulse sequence with t=10 sec
and an experiment recycle time of 10 sec, it is possible
to null and saturate all carbon intensity except that
originating from carbons having T_1~3 sec or less. Thus,
for the four spectra of drawn LPE shown in Figure 20, the
observed intensity originates almost exclusively from PE
chains in non-crystalline regions of the samples. (The
spectra have been normalized to represent the appropriate
amount of non-crystalline material in the sample.) A
distinguishing feature between the spectra is the relative
amount of aligned vs. misaligned chains. Clearly, the
two cold drawn samples (draw ratio 15) have a much higher
concentration of mobile chains which are aligned in the
draw direction (represented by the major resonance at
12.9 ppm) than the high temperature drawn sample (draw
ratio 12). This may be due to annealing processes which
can take place in the higher temperature drawing and
result in more perfect packing of chains and a concomitant
reduction in chain mobility. In the cold drawn samples,
larger concentrations of imperfections and voids may be
frozen-in, resulting in more volume for aligned chains
to undergo rotation about the chain axis, and thus
accounting for the relative intensities observed in the
spectra of Figure 20.

C. MAS of Polyethylene, cis-polybutadiene and Fluoropolymers

The DD/CP/MAS spectrum of semicrystalline PE[33] (ca. 60%
crystallinity) is shown in Figure 21 (top). It is obvious
that the resonance line is asymmetric on the high field
side. Observation of the spectra as a function of time
between 90° pulses using DD/MAS conditions shows the
presence of two peaks separated by ~2.3 ppm. It is
apparent the high field resonance line has a much shorter
carbon T_1 (having attained equilibrium intensity in 2s)
and a greater width (40-50Hz vs. 10Hz) than the low field
resonance line. These results are consistent with the
assignment of the low-field line to the crystalline
regions of PE and the high-field line to the
non-crystalline regions. In particular, the long T_1 of

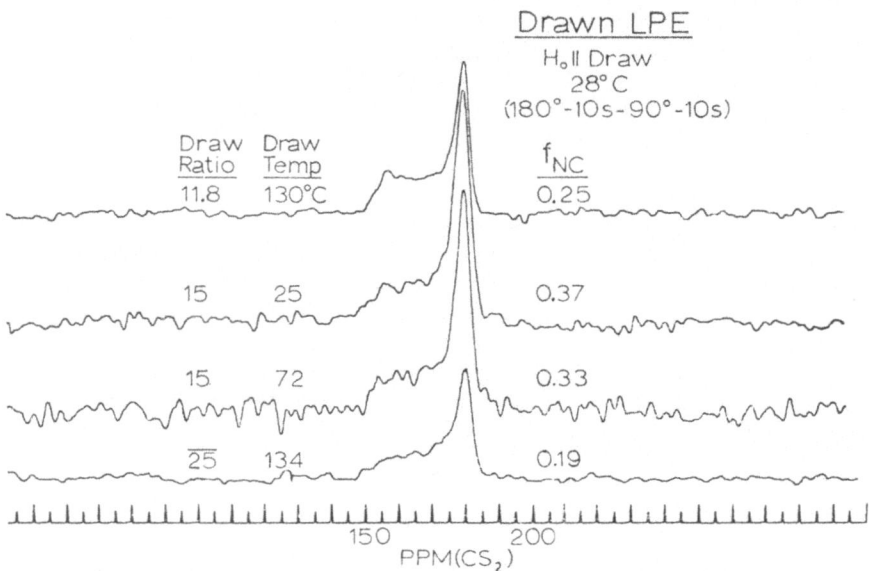

20. Proton-decoupled ^{13}C-NMR spectra from the
non-crystalline regions of linear polyethylene as a
function of draw ratio and temperature. The
non-crystalline fraction of material is indicated by f_{NC}.
The pulse sequence used is indicated in the figure.
Spectra are given on a scale normalized to f_{NC}. (Figure
from reference 59.)

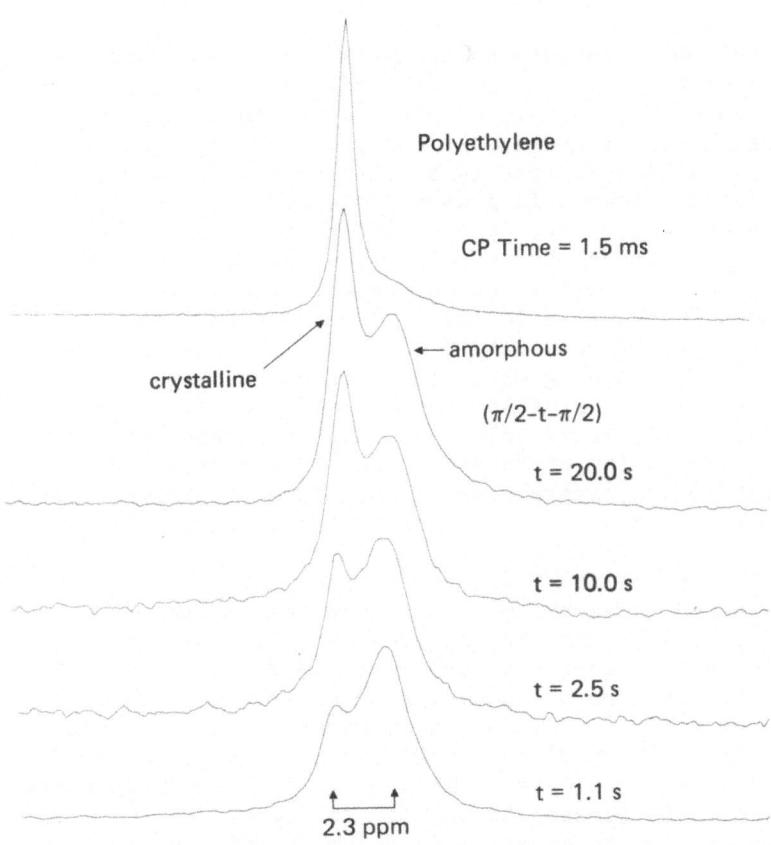

21. Magic-angle CP and π/2 ^{13}C-NMR spectra of
semicrystalline polyethylene at 27°C. The spectra
obtained with a π/2-t-π/2 pulse sequence used the
indicated values of t. The number of FID accumulations
varied from 2000-4500 but the spectra are displayed on a
normalized scale. The spectrum obtained from the CP
experiment used a CP contact time of 1.5ms and an
experiment recycle time of 2s. The spectrum is the result
of 400 FID accumulations. (Figure from reference 33.)

the low-field line is consistent with the limited mobility
of chains in the crystalline regions relative to
non-crystalline regions.

The 10Hz linewidth of the peak from the crystalline
region of PE is in accord with the linewidths (ca. 15Hz)
observed for the highly crystalline aromatic polyesters
discussed earlier. The 40-50Hz width of the peak from
non-crystalline regions is similar to the 50-100Hz
linewidths observed in glassy systems.[4] One consequence
of the chain rigidity in crystalline regions may be a
substantial reduction of motions in the 10^4-10^5Hz region,
as well as in the MHz region, relative to that for
glasses. Motions in 10^4-10^5 frequency region do not
result in complete motional averaging but instead yield
a dipolar broadening which is not removable by
realistically achievable proton-decoupling fields nor
with currently achievable magic angle spinning speeds.
Thus, a reduction of these motions in the aromatic
polyesters and crystalline PE may, in part, account for
the narrower resonance lines relative to glassy systems.[33]

If a greater homogeneity of the local environment
experienced by each repeat unit can be associated with
crystal habit, then the high degree of crystallinity of
the aromatic polyesters may result in substantial
reduction of line-broadening from dispersions of isotropic
chemical shifts. Such dispersions are associated with
the steric (in vinyl polymers) and conformational
isomerism common in the glassy state.[4] If reduced shift
dispersion is indeed the primary source of the
line-narrowing in crystalline polymers relative to
glasses, it would support the proposal[4] that it may not
be possible to narrow lines in glassy polymers below
40-50Hz.

The chemical shift between the resonance maxima in
PE presumably arises from constraint to trans bond
conformations in the crystalline region as opposed to
trans/gauche conformations in the amorphous regions.
Assuming the chemical shift difference (6.4 ppm) between
the crystalline peak in PE and the resonance of
cyclohexane which exists in an all gauche conformation
can be attributed entirely to conformational effects,
Earl and VanderHart[61] have concluded that the 2.3 ppm
shift between resonances for the crystalline and

non-crystalline regions is 0.6 ppm less than would be expected for a fraction of gauche states based on thermal considerations.[62] Although speculative, constraints on non-crystalline chains biasing the equilibrium population of trans/gauche towards trans would account for the difference. Such a result would also be in accord with the linewidth observed in the non-crystalline region. Motional effects account for only about one-half the total width; chemical shift disperison contributes ca. another 50% to the width.[61] The dispersion can be attributed to variations in trans/gauche population ratios in various chains or chain segments. These initial results suggest more detailed investigations, including variable temperature studies, are warranted to determine the potential of solid state NMR in elucidating the nature of semi-crystalline PE.

From the discussion on PE it is apparent that the static as well as the magic-angle spinning double resonance experiments provide useful information. However, we have seen earlier (e.g., Figure 3b) that overlap of the shielding anisotropies usually precludes detailed analysis of orientational effects in multi-line spectra. Fortunately, because the chemical shift anisotropy is totally static in origin, sample spinning at a rate much less than the anisotropy is sufficient to completely displace the spectral density from zero frequency and remove the line-broadening due to this source.[63] Of course, sidebands are prominent in slow spinning experiments when the chemical shift anisotropies are large; however, the intensity pattern of the spinning sidebands reflect the magnitude of the shift anisotropy and by proper analysis this information can be recovered by measuring spectra at a few spinning frequencies.[63,64] The utility of the technique has been demonstrated by Stejskal et al. for PMMA.[63] An alternate approach to obtain chemical shift anisotropy data in complex systems is to isotopically enrich with ^{13}C at sites of interest. Enrichment levels are required that produce a S/N sufficient to reduce the signal intensity from non-enriched sites to the noise level yet do not introduce problems from ^{13}C-^{13}C dipolar broadening. In less complex systems, variable angle spinning can be used to determine the elements of the shielding tensor. For example, in the static ^{13}C spectrum of cis-1,4 polybutadiene (CPBD) at -150°C only two elements (Figure 22a) of the shielding

tensor $\hat{\sigma}$ are resolved. The third overlaps the -CH$_2$-
pattern, thus making it impossible to fully assign the
tensor. However, variable angle spinning allows the full
assignment to be made through the angular functional
dependence of the spectrum (see Figure 22b). At the
magic-angle (Figure 22a) the spectrum collapses to a line
with FWHM (full-width at half-maximum) of 50Hz. With
the sample spinning at an angle of 87°, the P$_2$(cosθ)
functional dependence predicts a spectrum, relative to
the static case, reduced in width by 0.492 and with the
principal elements of the shielding tensor reversed with
respect to the isotropic shift.[13] On this basis, the
principal values of $\hat{\sigma}$ calculated from the data of Figure 22b
are σ_{11}=236±4, σ_{22}=115±4, σ_{33}=35±4. Assuming a similar
orientation of the shielding tensor as in ethylene
(σ_{11}=236; σ_{22}=124; σ_{33}=27)[65], these elements correspond
to shielding approximately perpendicular (but in plane)
to the double bond direction, parallel to the bond
direction, and perpendicular to the plane of the bond
respectively. The isotropic shift obtained from these
values is 128.7 ppm as compared to 129.7 determined from
the spectrum in the elastomeric state and the MAS spectrum
at -150. (Note, in the MAS spectrum, both resonances
despite large anisotropy differences, have the same FWHM
and side-bands are absent from the spectrum demonstrating
the efficiency of the spinning device (described in the
Experimental section) at low temperature.) A strong
driving force for observing chemical shift anisotropies
in polymers is that the manner in which the shielding
tensor undergoes averaging may provide insight as to the
nature of motions (i.e., along chain axis, perpendicular
to chain axis, etc) that enter a polymer chain as a
function of temperature (and thereby physical state).

Fluoropolymers are subject to analysis by DD/CP/MAS
techniques by using the ^{19}F nuclei to cross-polarize the
[13]C spins. Figure 23 displays the ^{13}C spectrum of a
highly-crystalline teflon at -120°C. The spectrum of
the spinning sample is reduced ca. 5-fold in width relative
to non-spinning sample. A well-defined anisotropy pattern
was not obtained and this may be due to the large anisotropy
of the fluorine resonance[66] coupled with a marginal
decoupling field. The combination of these factors
results in incomplete removal of C-F dipolar interactions
and thus failure to define the shape of the anisotropy
pattern (expected to be a non-axially symmetric pattern).

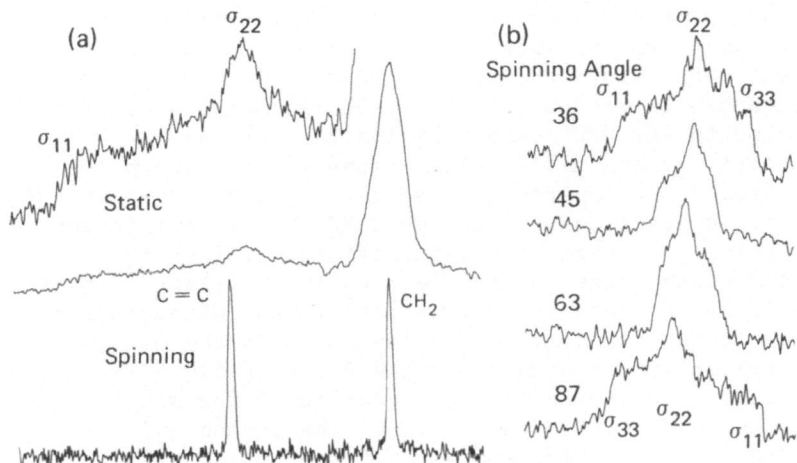

22.   a) DD/CP ^{13}C-NMR spectra of cis-1,4 polybutadiene at
-150°C - static and spinning at the magic-angle. For
the static spectrum, two elements of the shielding tensor
are indicated for the vinyl carbon in the ×4 vertical
expansion (inset). b) DD/CP spectra as a function of
spinning angle. Tensor elements are indicated.

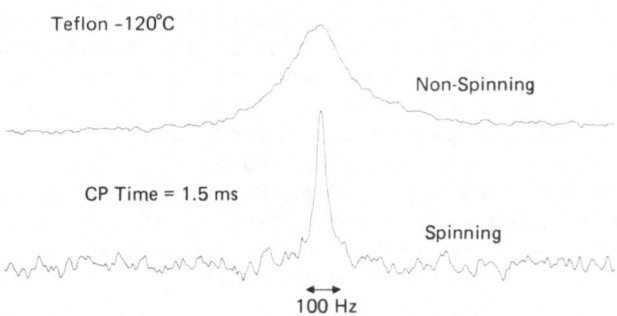

23.   Static and spinning fluorine-decoupled CP ^{13}C-NMR
spectra of highly crystalline poly(tetrafluoroethylene)
at -120°C.

The DD/CP/MAS spectrum of poly(chlorotrifluoroethylene) (CTFE) displays the expected two carbon resonances at ambient temperature and above (Figure 24). As the temperature is reduced, the upfield resonance, which is assigned to the $CFC\ell$ carbon begins to broaden substantially and ultimately disappears into the noise. This broadening apparently results from non-averaged $C-C\ell$ dipolar interactions. Although 3 KHz MAS is sufficient to partially average this interaction, the fact that $C\ell$ usually has a large quadrupole coupling constant[67] in an organic system such as CTFE results in the $C\ell$ nuclear moments being fixed at some orientation in the molecular axis system rather than along the Zeeman field. The moments follow the mechanical rotation of the sample to a degree dependent on the ratio of the Zeeman and quadrupole interaction energies of the nucleus,[68] resulting in line broadening and line splitting. At temperatures near T_g (~40), motion in the polymer backbone is apparently sufficient to cause the $C\ell$ T_1 to be short enough to induce "self-decoupling" of the $C-C\ell$ dipolar interaction.[69] At temperatures well below T_g motional averaging is severely reduced and thus the observed broadening.

D. Dynamics in the Solid State

In principal, resolution of individual carbon resonances in bulk polymers, allows relaxation experiments to be performed which can be interpreted in terms of main chain and side chain motions in the solid. In addition to the spin-lattice relaxation time in the Zeeman field, the spin-spin relaxation time and nuclear Overhauser enhancement, other parameters providing data on polymer dynamics include the proton and carbon spin-lattice relaxation times in the rotating-frame, $T_{1\rho}$, the cross-relaxation time T_{CR}, and proton relaxation in the dipolar field. Schaefer and Stejskal[4,70,71] have carried out pioneering work in exploring polymer dynamics using solid-state NMR techniques. Measurement of ^{13}C T_1 values in glassy PMMA at ambient temperature reveals that the $\alpha-CH_3$ carbon relaxes in <0.1s, the ester methyl and methylene carbons in ca. 1s and the two non-protonated (carbonyl and quaternary) carbons in ca. 10s. These results[71] are consistent with the onset of internal reorientation of $\alpha-CH_3$ at this temperature; relatively

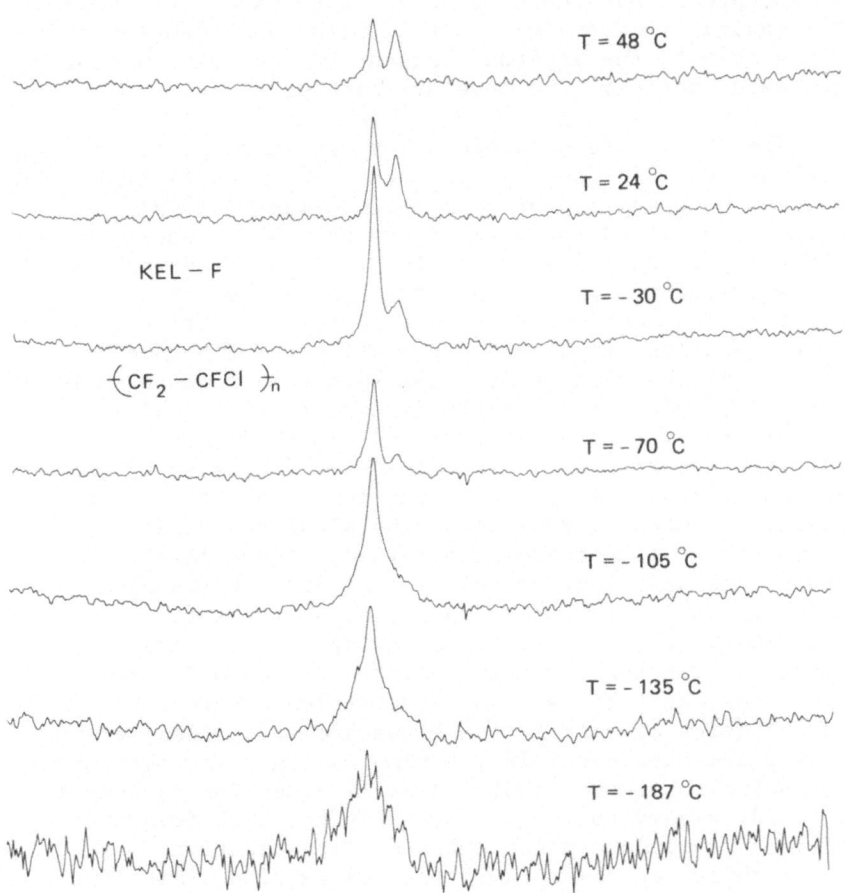

24. Fluorine-decoupled/CP/MAS ^{13}C spectra of PCTFE as
a function of temperature.

unhindered internal rotation of the side-chain methyl;
and restriction of any large-scale motion by the main chain
(T_g ~105°C). The presence of an approximately full
chemical shift anisotropy for the carbonyl carbon in the
non-spinning spectrum of PMMA at room temperature is a
direct proof that the ester side-group as a whole does
not engage in extensive internal rotational reorientations
comparable to the internal motions of the ester-methyl
(or ester-methoxy) and α-methyl carbons.

The NOE values determined for the three proton-bearing
carbons were all found to be about 2.0 which is less than
maximum (3.0)[37] but far from the minimum (1.15)[37] that
might be expected for a rigid system. This result argues
for the presence of a broad distribution of correlation
times associated with a variety of segmental rotational
and torsional motions. The presence of a distribution
of correlation times tends to level the NOE's of most
carbons to about the same value even though these carbons
may still have significantly different T_1's. Unlike the
situation for polymer systems well above T_g, the
distribution of correlation times[2,72-74] for solid PMMA
probably has a tail in the direction of short correlation
times. In other words, below the glass transition
temperature, long-range cooperative motions (necessarily
associated with long correlation times) are discriminated
against, while many short-range, high-frequency
(10^4-10^6Hz) torsional motions can apparently still
persist, even for the main chain, and even well below
T_g. Thus, deviations from a symmetrical, narrow
distribution of correlation times are most likely to be
toward the short-correlation time region. The net result
is that a substantial NOE can still exist for carbons in
a solid, glassy polymer. Naturally the high-frequency
motions resulting in the NOE are also effective
contributors to the spin-lattice relaxation times of the
various carbons. This is the reason the methylene-carbon
T_1 in PMMA has a short T_1 as compared to those found in
highly crystalline materials.

Long ^{13}C T_1's make measurement of T_1 and NOE by the
usual methods (e.g., 180°-t-90° sequence) tedious if not
prohibitative because of the long delay times required
for carbon re-polarization. Recently, Torchia[75] has
developed a method of determining T_1 which allows
repolarization by the more rapid CP process. Additionally,

the method appears to be well-suited for T_1 measurements in two-phase systems. Thus, it appears measurement of T_1 should be possible for most polymers; on the other hand, it is expected that the interpretation of the results will require considerable development of theory and models.

While the T_1 and NOE parameters are sensitive to polymer motions in the 15-100 MHz region, it is main chain motions in the 15-100 KHz region that are likely to be important in determining the mechanical properties of polymers below T_g. As described earlier, the KHz region of the frequency spectrum is accessible by measurement of the carbon rotating frame relaxation time, $T_{1\rho}^C$. Schaefer et al.[4] have measured this relaxation time for a number of glassy polymers using the experimental sequence depicted in Figure 25. Briefly, carbon magnetization is established by the Hartmann-Hahn CP procedure, contact between the ^1H and ^{13}C spin systems destroyed (i.e., H_{1H} turned-off), the carbons held in their rotating frame for a variable time, H_{1C} turned-off and the carbon magnetization sampled under dipolar decoupled conditions. $T_{1\rho}^C$ is then extracted from a semi-log plot of carbon intensity vs. hold time after breaking CP contact. Example spectra of a relaxation sequence for both spinning and non-spinning cases is given in Figure 26 for poly(phenylene oxide).

The interpretation of carbon $T_{1\rho}$ data is complicated by the fact that spin-spin (cross-relaxation) processes as well as rotating frame spin-lattice processes may contribute to the relaxation. This arises because the proton dipolar state is strongly coupled to the lattice. During the period in which carbon magnetization decays with H_{1H} off, carbon polarization can decay by motional processes ($T_{1\rho}^C$) or by polarization transfer to the proton dipolar state (T_{CR}), then to the lattice via the proton dipolar spin-lattice process (T_{1D}). Schaefer et al.,[70] VanderHart and Garroway[76] and Garroway et al.[49,77] have examined the problem in detail. While there remains some degree of controversy as to interpretation, at present, it seems that so long as the rotating-frame field is much greater than the proton local dipolar field, a system with motion in the KHz region will have $T_{1\rho}$ determined mainly by spin-lattice processes. According to Schaefer et al.[70] this will be the case for many glassy polymers

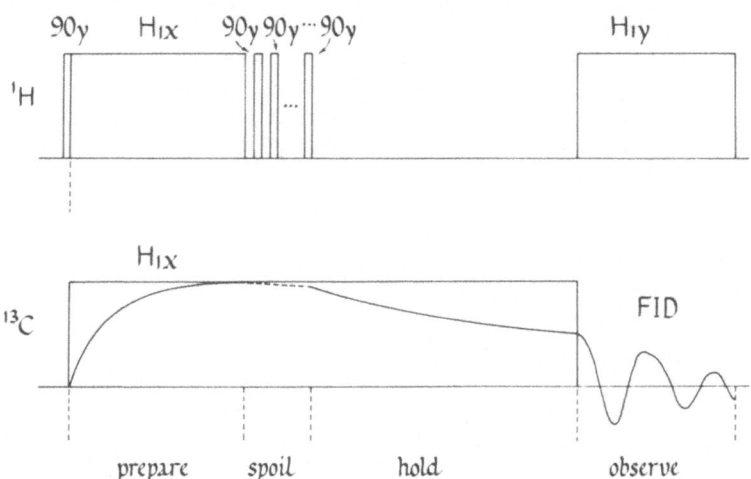

25. Pulse sequence for the ^{13}C $T_{1\rho}$ experiment. The first
part of the experiment is a matched Hartmann–Hahn CP
preparation of carbon magnetization. The second part
(the "spoil" sequence of 90° pulses separated by ^1H T_2's)
is usually omitted; it is included only when searching
for possible effects of transients in the ^1H channel on
the ^{13}C channel. The third part of the experiment is
the variable time the carbons are held in their rotating
frame without CP contact with the protons. This is
followed in the fourth part by the observation of the
carbon free induction decay with resonant dipolar
decoupling of the protons. (Figure from reference 4.)

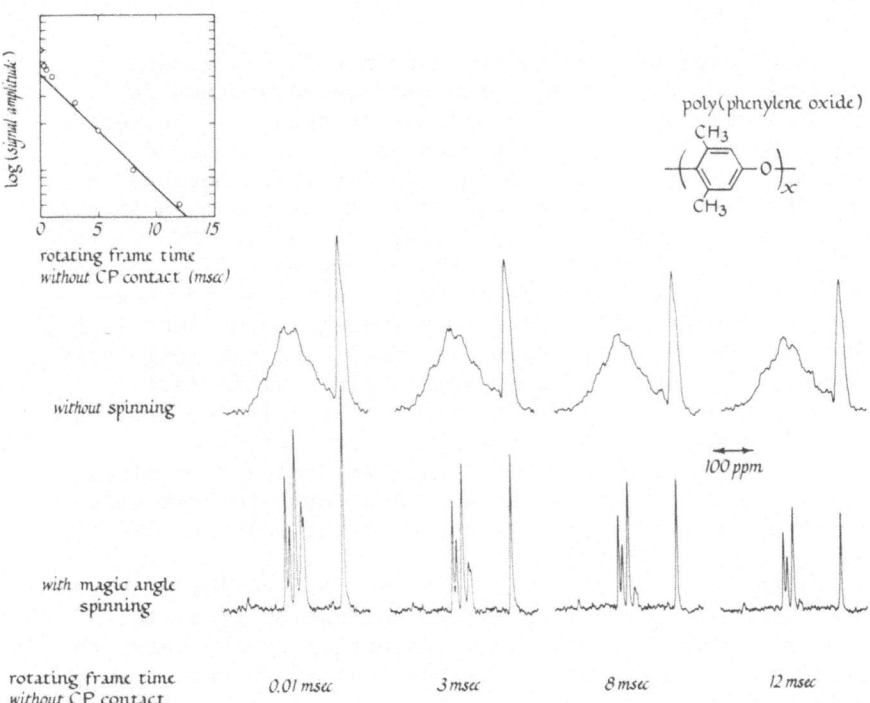

26. Cross-polarization ^{13}C-NMR spectra of poly(phenyleneoxide), with and without magic-angle spinning, as a function of the time the carbon magnetization was held in the rotating frame without CP contact. The insert shows the $T_{1\rho}^C$ relaxation behavior of the protonated aromatic-carbon doublet. (Figure from reference 4.)

and biopolymers at room temperature and probably plastic
crystals near the melting point. On the other hand,
crystalline polymers (for example PE) and small organic
molecules well below the melting point will show little
effect of spin-lattice processes on $T_{1\rho}^C$. It is possible
that heavily cross-linked polymers will form an
intermediate case.

   In a comprehensive paper on carbon $T_{1\rho}$ in glassy
systems, Schaefer et al. examined[4] seven systems in
considerable detail. The relaxation data was evaluated
in terms of spin-lattice processes and interpreted in
terms of distinct main chain and side chain motions in
the 10-50 KHz regime (note, for example, the much shorter
$T_{1\rho}^C$ value for the protonated ring carbons in PPO
(Figure 26) relative to the side-group methyl). The data
allowed conclusions to be drawn as to the short-range
nature of certain low-frequency side-group motions (e.g.,
the ester side-group motion in PMMA), and the long-range
cooperative nature of some main chain motions (e.g.,
polycarbonate and polysulfone). Interestingly, a direct
correlation was shown between the ratio of
cross-polarization and rotating frame relaxation times
($T_{CH}/T_{1\rho}^C$) for the main chain carbons of a polymer with
the toughness or impact strength of the polymer.

   Although much work remains to be done to fully assess
the value of the various carbon relaxation times in
understanding the mechanical properties of polymers, the
initial results appear promising even if ambiguous in
some instances.

                    E. Biopolymers

   Although in this paper we have concentrated on the
application of solid-state NMR techniques to synthetic
macromolecules, the techniques are also appropriate to
the study of biopolyers.[78] For example, Torchia et al.[79]
have utilized differences in the ^{13}C spectra of
deoxyhaemoglobin S obtained by scalar-decoupling, (proton)
dipolar decoupling, and cross-polarization to study the
intracellular gelation of deoxyhaemoglobin S. The gelation
is thought to be responsible for the "sickling"
deformation of the cell. Only isotropically mobile
haemoglobin molecules (correlation time, $\tau_c \lesssim 10^{-6}s$) are

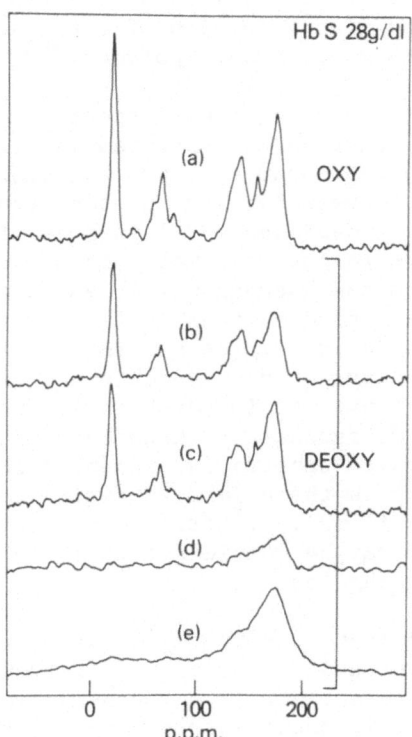

27. Comparison of spectra of a 28 g/dl preparation of
haemoglobin S at 37°C; a) oxygenated sample, 90°-t pulse
sequence, t=2s, scalar decoupled; b) deoxygenated sample,
90°-t pulse sequence, t=2s, scalar decoupled;
c) deoxygenated sample, 90°-t pulse sequence, t=2s, dipolar
decoupled; d) spectrum c) minus spectrum b); e) deoxygenated
sample, proton-enhanced, 1ms Hartmann-Hahn matched
contact, 2s repetition time dipolar decoupled. 4096
transients were accumulated in each case. Digital line
broadenings of 20Hz, in a), b) and c), and 50Hz, in d)
and e), were employed to enhance sensitivity. The
chemical shift scale is relative to external CS_2. (Figure
from reference 79.)

detected under scalar-decoupling conditions. By contrast,
dipolar-decoupling detects both mobile and motionally
restricted molecules. Comparison of the two types of
spectra thus allow discrimination of the fractional
amounts of deoxy-S monomer and "polymer."[80]

A typical sequence of decoupled spectra for
haemoglobin S in solution is presented in Figure 27.
Spectra 27a and b represent the oxy- and deoxy forms
obtained using low power (scalar) proton decoupling.
Clearly, there is a decrease (42%) in total intensity in
the deoxy relative to the oxygenated form - a result
which differs from the haemoglobin A case where the spectra
of both oxygenated and deoxygenated forms have the same
total intensity under scalar-decoupling conditions.[79]
In 27c is the spectrum of deoxy-S obtained using dipolar
decoupling conditions. An increase in intensity is
apparent and the difference spectrum between the scalar
and dipolar decoupled spectra is presented in 27d. The
residual intensity in the aliphatic region demonstrates
the presence of motionally restricted molecules in the
deoxy form, which may be considered to arise from gels
or haemoglobin aggregates.

The cross-polarized spectrum of deoxy-S, which
represents only intensity from carbons with non-vanishing
static dipolar interactions (i.e., non-isotropically
mobile molecules), is shown in 27e. Again, this result
is in contrast to the deoxy-A case where no CP spectrum
is observed. The width of the signal (ca. 150 ppm) observed
in the carbonyl-aromatic region of the CP spectrum is
typical of a chemical shift anisotropy for a crystalline
solid. Spin-lattice relaxation measurements indicate a
reduction of the backbone rotational correlation time by
10^4 upon gelation ($10^{-8} \rightarrow 10^{-4}$s). Thus, the set of results
is consistent with a two-phase model for the state of
deoxyhaemoglobin S in solution - a mobile monomeric phase
and an immobilized gel phase. Studies of haemoglobin
gelation in cells by the same NMR techniques indicates
as much as 80% of the haemoglobin is aggregated in
deoxygenated sickled erythrocytes.

Jelinski and Torchia[81] and Torchia and VanderHart[82]
have used double-resonance techniques to investigate the
mobility of the collagen peptide backbone in solution
and the fibrillar state. Specifically, collagen samples

TABLE 1

HR-NMR Parameters and the Features of Macromolecules

STRUCTURAL PARAMETERS

Chemical Shifts
Spin-spin Coupling Constants
Resonance Line Intensities

DYNAMICAL PARAMETERS

Resonance Linewidths
Spin-lattice Relaxation
    Times (T_1)
Spin-spin Relaxation
    Times (T_2)
Rotating Frame Relaxation
    Times ($T_{1\rho}$)
Nuclear Overhauser Factors

TYPES OF INFORMATION DERIVED
FROM LIQUID STATE SPECTRA

Primary structure of homopolymers, elucidation of chain
stereo-chemistry and conformation, elucidation of polymerization mechanisms, determination of rotational barriers,
determination of copolymer composition and sequence.

Determination of rates of
molecular reorientation and
rotational barriers, elucidation of chemical exchange
processes, data on the density
of high frequency (MHz) and low
frequency (KHz) segmental and
side chain motions in polymers.

in which glycine (every third residue in the chain) was
enriched to ca. 50% at either the carbonyl or the methylene
carbon were prepared and NMR parameters such as
linewidths, T_1 values, NOE and CP enhancements were
determined under a variety of physical conditions. The
results indicate that the collagen molecule in solution
undergoes torsional reorientation, as well as rod-like
reorientation about its long axis. The NMR data on the
glycyl residues provides strong evidence that rapid axial
motion ($\tau_c \sim 10^{-6}$) persists in the fibrillar state, however,
the angular range of reorientation is restricted to
ca. 30°.

V.   SUMMARY

The examples of ^{13}C-NMR spectra of bulk polymers
cited in Section IV clearly demonstrate that spectra of
solids in which resonance linewidths approach those of
liquids can be obtained by combining resolution (DD) and
sensitivity (CP) enhancement techniques with magic-angle
spinning. The results represent only a portion of
reported ^{13}C studies of bulk macromolecules. In addition
to the ^{13}C data, a considerable literature on
high-resolution proton NMR studies of solid polymers
exists and investigations utilizing ^{29}Si, ^{15}N, ^{31}P, ^{19}F
nuclei are beginning to be reported. The initial results
of high resolution NMR in solids suggest considerable
potential for DD/CP/MAS techniques in elucidating
structure and dynamics in the solid state. However, the
ultimate utility of these methods for defining the nature
of semi-crystalline polymers, etc. awaits the results of
much more investigation.

ACKNOWLEDGMENTS

The author would like to express his appreciation to
Dr. Jacob Schaefer (Monsanto Chemical Co., St. Louis, Mo.),
Dr. David VanderHart (NBS, Washington, D.C.),
Dr. Alan Garroway (NRL, Washington, D.C.), and
Dr. Dennis Torchia (NIH, Bethesda, Md.) for communicating
recent work in advance of publication and for helpful
discussions. Also, the author would like to express his
sincere appreciation to his colleagues, Dr. Colin Fyfe
(Univ. of Guelph, Guelph, Ontario, Canada) and

Dr. Costantino Yannoni (IBM, San Jose, Ca.) for their
painstaking efforts in introducing the nuances of solid
NMR and for their many valuable discussions during the
course of our collaborative efforts. The author would
like to acknowledge the inventive technical assistance
of R. A. Kendrick in modifying the commercial spectrometer
and adding new "twists" to the experiments. Finally,
the author would like to thank Virginia Armetta for her
expert help in preparing this manuscript.

REFERENCES AND NOTES

1. I. R. Peat and W. F. Reynolds, Tetrahedron Lett. 14,
    1539 (1972).

2. J. R. Lyerla, Jr., T. T. Horikawa and D. E. Johnson,
    J. Amer. Chem. Soc. 99, 2463 (1977).

3. For a recent review see V. J. McBrierty, Polymer 15,
    503 (1974).

4. J. Schaefer, E. O. Stejskal and R. Buchdahl,
    Macromolecules 10, 384 (1977), and references therein.

5. E. R. Andrew in "Progress in Nuclear Magnetic
    Resonance Spectroscopy," J. W. Emsley, J. Feeney and
    L. H. Sutcliffe, Eds., Pergamon Press, New York, N.Y.,
    1972, Vol. 8, Chapter 1.

6. Although the symbols I and S appear in some equations
    as spin operators or spin quantum numbers, their
    meaning in these instances is obvious and no ambiguity
    results.

7. C. P. Slichter, "Principles of Magnetic Resonance,"
    Springer-Verlag, New York, N.Y., 1978, p. 71. Because
    S is a "rare" spin, we neglect the $\langle\Delta\omega^2\rangle_{SS}$
    contribution.

8. For example, methyl group reorientation about the
    C_3 axis.

9. Reference 7, Chapter 3.

10. B. Schneider, D. Doskocilová and H. Pivocá, "Magnetic
    Resonance in Chemistry and Biology," Proceedings
    International Summer School, Yugoslavia, 1971. Note
    the definitions of α and β are interchanged here to
    be consistent with use of β in Eq. 6.

11. K. W. Zilm, D. W. Alderman and D. M. Grant, J.
    Magnetic Resonance 30, 563 (1978). These authors
    cite that rates of 21 KHz have been obtained by
    J. W. Beams (Rev. Sci. Instrum. 8, 795 (1937)) using
    H_2 as the driving gas.

12. F. Bloch, Phys. Rev. 111, 841 (1958).

13. See, for example, G. Sinning, M. Mehring and A. Pines, Chem. Phys. Lett. 43, 382 (1976) and references therein. The I-I flip-flop term, if non-zero, results in the necessity of having the I decoupling field large compared to the flip-flop rate for I spins to decouple the I-S interaction. See, M. Mehring, NMR: Basic Principles and Progress, Vol. 11, Springer-Verlag, New York, 1976, p. 153.

14. A. Pines, M. G. Gibby and J. S. Waugh, J. Chem. Phys. 59, 569 (1973).

15. A. Pines, M. G. Gibby and J. S. Waugh, Chem. Phys. Lett. 15, 373 (1972).

16. J. Schaefer in "Structural Studies of Macromolecules by Spectroscopic Methods," K. J. Ivin, Editor, John Wiley and Sons, New York, N.Y., 1976, Chapter 11.

17. A. Abragam, "The Principles of Nuclear Magnetism," Oxford Press, Oxford, 1961, Chapter IX.

18. D. L. VanderHart, National Bureau of Standards. Private communication.

19. N. Bloembergen, "Nuclear Magnetic Relaxation, W. A. Benjamin, Inc. New York, N.Y., 1961.

20. The exchange term is contained in the $\vec{I}_i \cdot \vec{I}_j$ and $\vec{I}_k \cdot \vec{I}_m$ terms of the Hamiltonians given in Eqs. 4b and 4d.

21. M. Goldman, "Spin Temperature and NMR in Solids," Oxford University Press, Oxford, 1970, Chapter 3.

22. These arguments are simplistic; clearly consideration must be given to how "good" the thermal contact is between the two regions (i.e., the size of the interfacial regions) whether all the crystalline region makes the same thermal contact with the amorphous region (size of the crystallites is important), etc.

23. Reference 17, page 2.

24. T. M. Connor, in NMR Basic Principles and Progress, Vol. 4, Springer-Verlag, New York, N.Y., 1971, p. 247-269.

25. To describe the motion of spins under the influence of a resonant rf field, H_1, it is conventional in magnetic resonance to transform to a rotating frame representation. In the rotating frame, the spins are subject to an effective field, $\vec{H}_{eff}$,

$$\vec{H}_{eff} = (H_0 - \omega/\gamma)\hat{z} + H_1\hat{x}$$

where $\hat{x}$ and $\hat{z}$ are unit vectors along the x and z axis of the rotating frame (note, the z axis is coincident with the laboratory z-direction) and ω is the frequency of H_1. Thus in the rotating frame representation, there exists two spin energy levels (I=1/2) with Zeeman splitting

$$\omega_{eff} = \gamma H_{eff}$$

At resonance, $(H_0 - \omega/\gamma) = 0$, and thus the magnetization precesses about $\vec{H}_1$. More detail can be found in texts such as T. C. Farrar and E. D. Becker, "Pulse and Fourier Transform NMR, Academic Press, New York, N.Y., Chapters 1-3.

26. The terms involving $M_1^I(\infty)$, the actual equilibrium intensity, are neglected as $M_1^I(\infty)$ is too small to be observed.

27. Viewed from the rotating frame, the relaxation is governed by T_1-like processes even though in the laboratory frame it is a relaxation of transverse magnetization, a T_2-like process.

28. The population distribution between the two spin states is conveniently described by the concept of a spin temperature. For a Boltzmann distribution $T_S = T_L$; however, for a non-Boltzmann distribution $T_S \neq T_L$ but approaches T_L in the spin-lattice relaxation time. See Ref. 21 for details.

29. S. R. Hartmann and E. L. Hahn, Phys. Rev. 128, 2042 (1962).

30. Note, the secular part of the IS dipolar Hamiltonian
    (Eq. 4c) contains no flip-flop term, as appears in
    Eqs. 4b and 4d, due to the frequency mis-match of I-
    and S-spin systems in H_0; thus, the effect of the
    Hartmann-Hahn experiment is to make the IS flip-flop
    term secular.

31. J. Schaefer, E. O. Stejskal and R. Buchdahl,
    Macromolecules $\underline{8}$, 291 (1975).

32. J. Schaefer and E. O. Stejskal, J. Amer. Chem. Soc.
    $\underline{98}$, 1031 (1976).

33. C. A. Fyfe, J. R. Lyerla, W. Volksen and C. S. Yannoni,
    Macromolecules $\underline{12}$, 000 (1979).

34. C. A. Fyfe, J. R. Lyerla and C. S. Yannoni, J. Amer.
    Chem. Soc. $\underline{100}$, 5635 (1978).

35. C. A. Fyfe, H. Mossbrugger and C. S. Yannoni, J.
    Magn. Resonance, in press (1979).

36. M. E. Stoll, A. J. Vega and R. W. Vaughan, Rev. Sci.
    Instrum. $\underline{48}$, 800 (1977).

37. From the purely operational viewpoint, the existence
    of an NOE provides increased S/N. However, the
    two-level decoupling scheme also allows measurement
    of the NOE and thus additional insight into the
    molecular dynamics of the system under investigation.
    (For a discussion of the $^{13}C-\{^1H\}$ NOE see, J. R. Lyerla
    and G. C. Levy, "Topics in Carbon-13 NMR Spectroscopy,
    J. Wiley and Sons, Interscience, New York, N.Y. 1974,
    Vol. 1, Chapter 3.)

38. E. O. Stejskal and J. Schaefer, J. Magn. Resonance
    $\underline{18}$, 560 (1975).

39. I. J. Lowe, Phys. Rev. Letters $\underline{2}$, 285 (1959).

40. At this angle, the amplitude of the perpendicular
    component of H_{1I} is 0.816 H_{1I}. This results in a
    voltage induced by transverse magnetization reduced
    by the same factor. The loss in S/N is compensated
    (under MAS conditions) by the better filling factor
    relative to a perpendicular arrangement.

41. The desired ratio of $T_{1\rho}^I/T_1^I$ being as close to unity as possible. Of course, account must be taken of the other factors which affect the recycle time such as the time required to collect data for the desired resolution (i.e., digital resolution), etc.

42. Despite numerous start-ups and "crashes," the Kel-F and Delrin rotors exhibit little wear after several hundred hours of operation.

43. Determined from the maximum linear velocity of sound in He and the diameter of the rotor periphery $(\omega=(v^2/r^2)^{1/2})$.

44. R. G. Pembleton, L. M. Ryan and B. C. Gerstein, Rev. Sci. Instrum. $\underline{48}$, 1286 (1977).

45. Chemical and Engineering News, October 16, 1978, p. 23.

46. J. Economy, R. S. Storm, V. I. Matkovich, S. G. Cottis and B. E. Nowak, J. Polym. Sci., Polym. Chem. Ed. $\underline{14}$, 2207 (1976).

47. Also, the number of interactions is important, as well as effects of molecular motion. For example, a methyl group would cross-polarize more rapidly than a methylene or methine in a rigid-lattice situation; however, if the methyl group undergoes rapid reorientation about the C_3-axis, partial averaging of the dipolar interaction will occur which may result in the methyl carbon being polarized more slowly than methylene or methine carbons despite the greater number of C-H interactions.

48. J. B. Stothers, "Carbon-13 NMR Spectroscopy," Academic Press, New York, N.Y., 1972, p. 197.

49. A. N. Garroway, W. B. Moniz and H. A. Resing, ACS Symposium Series (1978). To be published.

50. W. J. Jackson, Jr. and H. F. Kuhfuss, J. Polym. Sci. Polym. Chem. Ed. $\underline{14}$, 2043 (1976).

51. A. Blumstein, "Liquid Crystalline Order in Polymers," Academic Press, New York, N.Y., 1978.

52. G. R. Luckhurst, in "Liquid Crystals and Plastic
    Crystals," G. W. Gray and P. A. Winsor, Editors, Ellis
    Horwood Ltd., Chichester, U.K., Vol. II, Chpt. 7.

53. A. Pines and J. J. Chang, Phys. Rev. A10, 946 (1974).

54. A. Pines and J. J. Chang, J. Am. Chem. Soc. 96, 5590
    (1974).

55. A. Pines, D. J. Rueben and S. Allison, Phys. Rev.
    Lett. 33, 1002 (1974).

56. J. Urbino and J. S. Waugh, Proc. Nat. Acad. Sci. USA
    71, 5062 (1974).

57. D. L. VanderHart, J. Chem. Phys. 64, 830 (1976).

58. D. L. VanderHart, J. Magn. Resonance 24, 467 (1976).

59. D. L. VanderHart, Macromolecules, in press.

60. R. A. Komoroski, J. Maxfield, F. Sakaguchi and
    L. Mandelkern, Macromolecules 10, 550 (1977).

61. W. Earl and D. L. VanderHart, Macromolecules 12, 000
    (1979).

62. At 300°K and using a barrier of 600 cal for t-g
    conformational jump, a fraction of gauche states
    equal to 0.42 is calculated.

63. E. O. Stejskal, J. Schaefer and R. A. McKay, J. Magn.
    Resonance 25, 569 (1977).

64. E. Lippmaa, M. Alla and T. Tuherm, in Magnetic
    Resonance and Related Phenomena, H. Brunner,
    K. H. Hausser and D. Schweitzer, Heidelberg-Geneva,
    1976.

65. K. W. Zilm, R. T. Conlin, D. M. Grant and J. Michl,
    J. Am. Chem. Soc. 100, 3038 (1978).

66. M. Mehring, R. G. Griffin and J. S. Waugh, J. Chem.
    Phys. 55, 746 (1971).

67. B. P. Dailey, J. Chem. Phys. <u>33</u>, 1641 (1960)

68. E. Lippmaa, M. Alla and E. Kundla, presented at the
    18th Experimental NMR Conference, Asilomar, Ca.,
    April 1977.

69. Reference 17, Chapter VIII.

70. E. O. Stejskal, J. Schaefer and T. R. Steger, J.
    Magn. Resonance, in press.

71. J. Schaefer in "Structural Studies of Macromolecules
    by Spectroscopic Methods," K. J. Ivin, Ed.,
    Wiley-Interscience, New York, N.Y., 1976, Chpt. 11.

72. F. Heatley and A. Begum, Polymer <u>17</u>, 399 (1976).

73. J. Schaefer, Macromolecules <u>6</u>, 882 (1973).

74. J. R. Lyerla, Jr. and D. A. Torchia, Biochemistry
    <u>14</u>, 5175 (1975).

75. D. A. Torchia, J. Magn. Resonance <u>30</u>, 613 (1978).

76. D. L. VanderHart and A. N. Garroway, submitted to J.
    Magn. Resonance (1979).

77. A. N. Garroway, W. B. Moniz and H. A. Resing, to be
    published in Faraday Soc. Symposium 13 (Dec. 1978).

78. D. A. Torchia and D. L. VanderHart, Topics in
    Carbon-13 NMR Spectrosc. <u>3</u>, in press.

79. J. W. H. Sutherland, W. Egan, A. N. Schechter and
    D. A. Torchia, Biochemistry, in press (1979).

80. In this context, the word polymer refers to an
    aggregated or highly associated state, not a long
    chain molecule with covalent bonds.

81. L. W. Jelinski and D. A. Torchia, J. Mol. Biol., in
    press (1979).

82. D. A. Torchia and D. L. VanderHart, J. Mol. Biol.
    <u>104</u>, 315 (1976).

83. Teflon and Delrin are registered trademarks of the
    E. I. duPont Nemours & Co. (Inc.); Lexan is a registered
    trademark of the General Electric Co; Kel-F is a
    registered trademark of the 3M Co.

SURFACE STUDIES ON MULTICOMPONENT POLYMERIC SOLIDS

James J. O'Malley and H. Ronald Thomas

Xerox Corporation, Webster Research Center

800 Phillips Road, Webster, New York   14580

INTRODUCTION

Block copolymers and polyblends frequently exhibit phase separation which typically gives rise to a dispersed phase consisting of one polymer component in a continuous matrix of the second polymeric component. The detailed morphology of the domain structure in these systems depends upon such factors as the relative proportions of the two components, the molecular weights, the thermal and physical histories of the polymers, and for solvent cast films, upon the solvent and temperature.

A variety of experimental techniques have been employed in investigations of the domain structure of multicomponent polymeric solids with the predominant emphasis being on the bulk morphology. By contrast, the detailed structure of the surface, i.e., the outermost few tens of angstroms, of such polymer systems has been subject to much less detailed consideration. Since many properties of a polymeric solid depend on the detailed structure of the surface, however, and since the latter may be considerably different from the bulk, techniques which can, in principle, differentiate surface from bulk properties are likely to be of considerable interest in this area.

In this paper we will explore the application of the relatively new technique of X-ray photoelectron spectroscopy (XPS) to the quantitative evaluation of the surface structure of multicomponent polymeric solids. This paper

is divided into five sections. In the first and second
sections we describe the XPS experiment and its initial
application in the study of the surface properties of
polystyrene-polydimethyl siloxane block copolymers. In the
third section of the paper, we describe angular XPS studies
[XPS (θ)] and this is followed by a section on polystyrene/
polyethylene oxide block copolymers and the utilization of
the XPS (θ) technique in depth profiling the compositional
variations near the polymer-air interface. Our concluding
remarks are contained in the final section of the paper.

THE XPS EXPERIMENT

The essential features of an electron spectrometer
are shown in Figure 1. Simply, the XPS experiment involves
the measurement of binding energies of electrons in molecules by determining the energies of electrons ejected by
the interactions of a molecule with a monoenergetic beam of
X-rays.[1,2] $MgK\alpha_{1,2}$ with a photon energy of 1253.7eV is

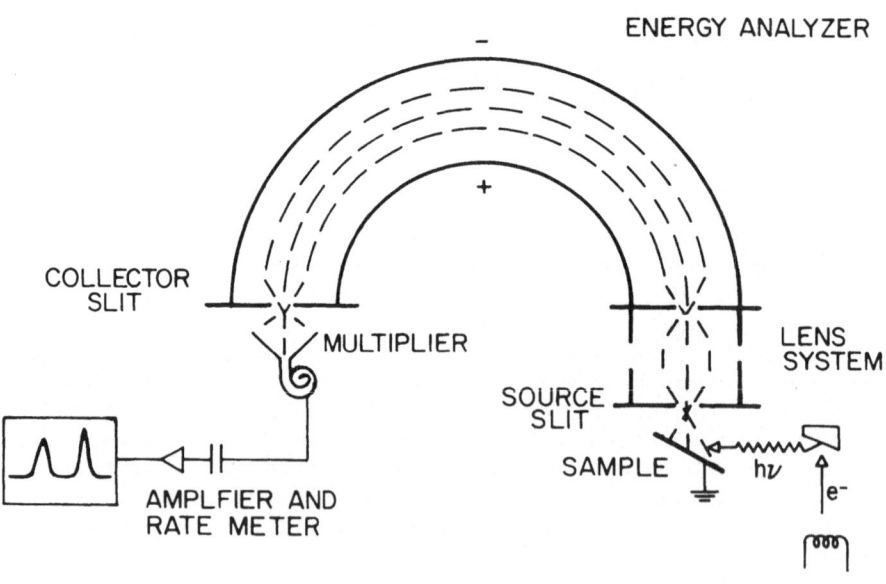

FIGURE 1

Schematic of X-ray photoelectron spectrometer

commonly used as the X-ray source. With a knowledge of the
kinetic energy of the photon source (hν) and the measured
kinetic energy (K.E.) of the photoemitted electron, one can
calculate the binding energy (B.E.) according to Equation 1.

$$B.E. = h\nu - K.E. \qquad (1)$$

The calculated binding energy is the energy required to
remove an electron to infinity with zero kinetic energy
and its absolute value is characteristic of the atom from
which the core electron has been photoemitted.

The removal of a core electron is accompanied by substantial reorganization of the valence electrons in response
to the effective increase in nuclear charge. This perturbation gives rise to a finite probability that photoionization
will be accompanied by simultaneous excitation of a valence
electron from an occupied to an unoccupied orbital [$\pi^* \leftarrow \pi$].
These relatively weak shake-up transitions are observed as
satellite peaks in the XPS spectrum, and they are found on
the low kinetic energy side of the main photoionization
peak. These shake-up peaks add an extra dimension in the
analysis of XPS data[3-5] and they will be used extensively
in the analysis of the block copolymers.

POLYSTYRENE-POLYDIMETHYL SILOXANE BLOCK COPOLYMERS

The initial literature report on the application of
XPS to the analysis of the surface properties of block copolymers appeared only recently.[6] XPS was used to characterize the surface compositions of two polystyrene (PS)-
polydimethyl siloxane (PDMS) diblock copolymers, one
containing 23 wt. % PS ($\overline{M}_n$ = 121,000) and the other containing 59 wt. % PS ($\overline{M}_n$ = 124,000), and to determine the
influence of film casting solvent on the composition at the
air-polymer interface. In Figure 2 are shown representative
XPS spectra of thin films of PS, PDMS and the 59 wt. % PS
copolymer cast from cyclohexane and styrene, preferential
solvents for PDMS and PS, respectively. The signals from
the O_{1s} and Si_{2p} core levels in PDMS are clearly evident
as is the C_{1s} core level signal that emanates from both
the PS and PDMS components. The spectrum for PS homopolymers
reveals the presence of a low intensity shake-up transition
($\pi^* \leftarrow \pi$) as a satellite of the main C_{1s} photoionization
peak. This shake-up transition, characteristic of PS, is
also found in the spectrum of the copolymer cast from

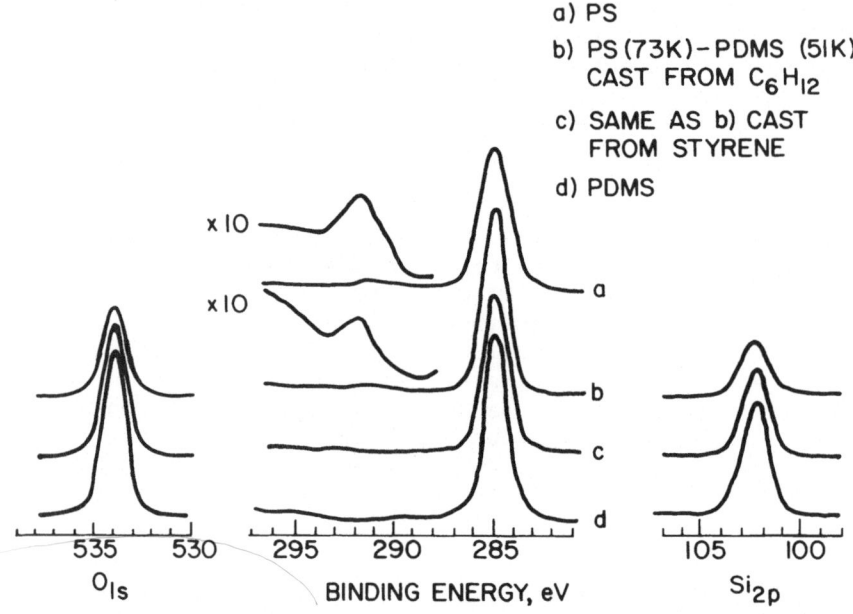

a) PS

b) PS(73K)-PDMS (51K)
   CAST FROM C_6H_{12}

c) SAME AS b) CAST
   FROM STYRENE

d) PDMS

FIGURE 2

XPS spectra of polymer films showing the changes in the
intensity ratios of the core electron signals and the
appearance of the shake-up satellites: a) Polystyrene;
b) 59 wt. % PS copolymer cast from cyclohexane; c) 59 wt.
% PS copolymer cast from styrene; d) polydimethylsiloxane.

cyclohexane. The absence of this transition from the spectrum of the copolymer cast from styrene immediately permits
the qualitative assessment that the cyclohexane cast copolymer film contains more polystyrene at the air-polymer interface than does the styrene cast copolymer film.

The experimental intensity ratios of the C_{1s}/Si_{2p},
C_{1s}/O_{1s} and O_{1s}/Si_{2p} peaks are tabulated in Table 1.

TABLE 1

Experimental Intensity Ratios in the XPS Spectra of Films
of PS, PDMS, and PS-PDMS Block Copolymers

Polymer Film	$I(C_{1s}/Si_{2p})$	$I(C_{1s}/O_{1s})$	$I(O_{1s}/Si_{2p})$
PDMS (I^{∞} values)	1.66	1.20	1.38
PS	--	--	--
PS-PDMS (23% PS) Diblock	1.7	1.3	1.35
PS-PDMS (59% PS) Diblock			
Cast from styrene	1.77	1.28	1.38
Cast from cyclo-hexane	3.1	2.05	1.5

It is evident from these data that the 23 wt. % PS
copolymer and the film of the 59 wt. % PS copolymer cast
from styrene are indistinguishable from pure PDMS by the
XPS technique, thus indicating that in these samples the
PDMS component forms the surface and immediate subsurface.
With a knowledge of the appropriate electron escape depths
as a function of kinetic energy,[7,8] we can estimate the
PDMS surface layers to be at least 40Å deep in these films.
This is consistent with the fact that shake-up satellites
attributable to the PS component are absent in the spectra
of these films. On the other hand, the intensity ratios
for the 59 wt. % PS copolymer cast from cyclohexane indicate
that a small amount of PS is being detected with the PDMS
component. Contact angle measurements,[6] which showed
that all the above films had critical surface tensions ~22

dyne/cm, a value typical of PDMS, suggest, however, that
the first monolayer at the air-polymer interface corresponds to an essentially pure PDMS component.

If it is assumed, then, that the copolymer films
consist of an outer layer of pure PDMS, the intensity ratios
of the various core levels determined by XPS can be used to
estimate the thickness of this layer. As a simple model,
suppose each film consists of a continuous layer of PDMS
of thickness d covering the polymer bulk, which is composed
of 59 wt. % PS and 41 wt. % PDMS. By using the experimentally determined intensity ratios listed in Table 1 along
with the electron escape depths,[3] λ, for C_{1s} = 10Å,
O_{1s} = 8Å and Si_{2p} = 11.5Å, three independent estimates of
d are available by substitution into Equations 2 and 3,

$$I_o = I_o^\infty (1-e^{-d/\lambda}) \qquad (2)$$

$$I_s = I_s^\infty (e^{-d/\lambda}) \qquad (3)$$

where I_o and I_s are the intensities of the signals from a
given core level arising from the surface overlayer (I_o) and
the subsurface of bulk (I_s) and I^∞ is the intensity observed
for an infinitely thick layer.[6] The calculated PDMS surface layer thicknesses for the block copolymers are shown
in Table 2.

TABLE 2

Calculated Thickness of the Surface PDMS Layer in PS-PDMS
Diblock Copolymers

	Calculated Thickness of PDMS Surface (Å)		
Copolymer Film	From $I(C_{1s}/Si_{2p})$	From $I(C_{1s}/O_{1s})$	From $I(O_{1s}/Si_{2p})$
23 wt. % PS Copol.	>40	>40	>40
59 wt. % PS Copol. Cast from styrene	>40	>40	>40
Cast from cyclo- hexane	13	13	13

The consistency in the values obtained for a given copolymer film can be taken as an indication of the reliability of the treatment and the constants used. The predominance of the PDMS component in the surface undoubtedly reflects the substantially lower surface free energy of PDMS compared with PS.

DEPTH PROFILING THE SURFACES OF POLYMERS

Up to this point, all the XPS measurements that we have discussed have been made by analyzing the photoemitted electrons normal to the surface of the sample. This experimental arrangement is the most commonly used in XPS studies, and with it the effective sampling depth is maximized (typically top 50Å). We can further refine the XPS measurements by controllably decreasing the effective sampling depth via angular measurements, XPS (θ). The XPS technique is shown schematically in Figure 3.

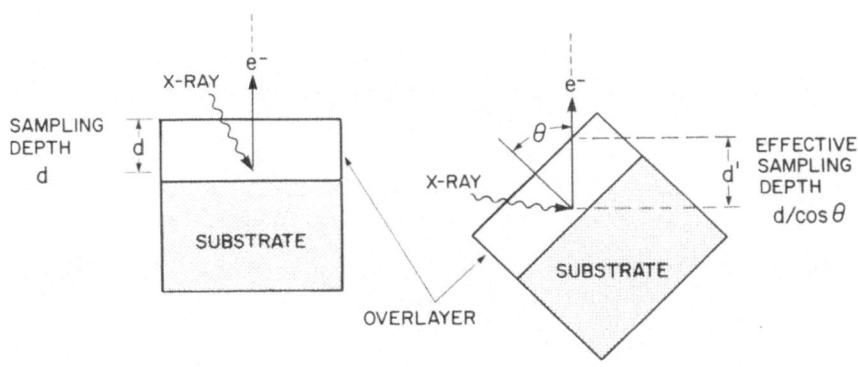

FIGURE 3

Schematic of XPS (θ) measurement technique. Sample is rotated and spectra are recorded as a function of electron take-off angle θ.

The sample is rotated relative to the fixed position energy analyzer by angle θ, designated as the angle between the normal to the sample and the slits in the analyzer. It is readily seen that electrons collected at grazing exit angles (θ → 90°) relative to the surface will enhance surface features relative to electrons collected normal to the surface.[2,9,10] This, in effect, allows for depth profiling the composition of the upper 50Å of the surface and provides us with a means to explore the molecular organization of the surface macromolecules.

POLYSTYRENE–POLYETHYLENE OXIDE BLOCK COPOLYMERS

In this section we will describe the results of our angular dependent XPS measurements, XPS (θ), which were made on a series of polystyrene (PS)-polyethylene oxide (PEO) diblock (PS-PEO) and triblock (PEO-PS-PEO) copolymers. Table 3 contains the copolymer composition and molecular weight data for the various block copolymers employed in this study. The synthesis, purification and characterization of these materials has been described previously.[11]

TABLE 3

Characterization Data for PS-PEO and PEO-PS-PEO Block Copolymers

	% PS		Segment $\overline{M}_n$ $(x10^{-3})$	
Samples	Wt.	Mole	PS - PEO	PEO - PS - PEO
Diblocks	19.6	9.6	21.2 – 87.2	
	39.3	21.4	13.2 – 20.3	
	70.0	49.5	24.3 – 10.4	
Triblocks	23.5	11.4		8.5 – 5.1 – 8.5
	38.5	21.0		8.4 – 10.5 – 8.4
	70.3	49.8		9.1 – 43.2 – 9.1

XPS core level spectra for PS, PEO and a representative PS-PEO diblock copolymer are shown in Figure 4. Thin polymer films cast from chloroform solution were used for these measurements. The PS spectrum has a strong peak centered at 285eV associated with the direct photoionization of the C_{1s} core levels and a low energy satellite peak at 291.6eV arising from the $\pi^* \leftarrow \pi$ shake-up

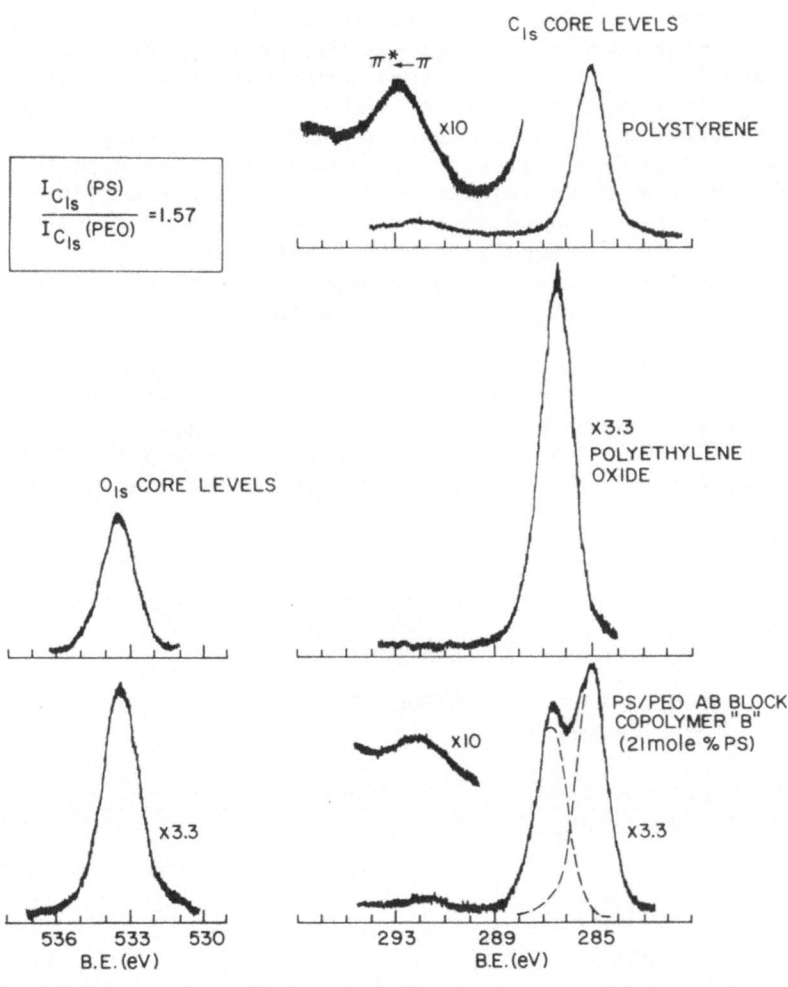

FIGURE 4

XPS spectra of polystyrene (top), polyethylene oxide (middle) and a polystyrene-polyethylene oxide diblock copolymer (bottom).

transition.   The spectrum for PEO shows a single peak for the C_{1s} levels at 286.5eV and a single peak for the O_{1s} core levels at 533.3eV.   The peak area ratio of the C_{1s} to O_{1s} core levels is 0.73.   The spectrum of the PS-PEO block co-polymer shows the characteristic O_{1s} peak of PEO, the shake-up satellite of PS, and an easily deconvoluted doublet for the C_{1s} core levels in PS and PEO.   Comparing the PS and PEO homopolymer spectra, we find that the C_{1s} core level peak in PS is 1.57 times larger than the C_{1s} peak in PEO. Applying this intensity ratio of 1.57 as a correction factor to the measured C_{1s} peak areas for the PS and PEO components in the copolymers, it is straightforward to calculate the surface composition of the copolymers from the C_{1s} (PS)/C_{1s} (PEO) intensity ratio.

The diblock and triblock copolymers in Table 3 were cast into films from chloroform solutions and their XPS spectra were recorded at three different angles, $\theta = 0°$, 45° and 80°, in order to achieve effective sampling depths of ~50, 25 and 10Å, respectively.   The calculated surface compositions at different sampling depths are shown in Table 4.

TABLE 4

Surface Compositions of PS/PEO Block Copolymer Films Determined by the XPS (θ) Technique

Bulk Comp.	Copolymer	Surface Comp. (Mole % PS)		
(Mole % PS)	Type	$\theta = 0°$	$\theta = 45°$	$\theta = 80°$
~10	Diblock	42	54	62
	Triblock	39	45	63
~20	Diblock	46	52	56
	Triblock	58	62	68
~50	Diblock	82	85	86
	Triblock	84	87	89

The data clearly reveals that both components in the copoly-mer are present at the surface and that there is a consider-able surface excess of PS compared to the bulk composition.

Both the diblock and triblock copolymer samples exhibit
similarly large surface excesses of PS and the PS
concentration in the surface region gradually increases as
the air-polymer interface (θ = 80°) is approached. This is
consistent with contact angle measurements which indicate
that PS has a critical surface tension of 36 dynes/cm
compared to a value of 44 dynes/cm for PEO.

Various models for the surface topography of these
block copolymers have been considered.[12,13] A continuous
overlayer model, as described by Equations 2 and 3 and used
to analyze the PS-PDMS block copolymer data, is not applicable to the PS/PEO copolymers. In this model, PS would
comprise the continuous overlayer because there is a surface
excess of PS in these copolymers and because it has a lower
critical surface tension than PEO. This model predicts,
among other things, that the C_{1s} (PEO)/O_{1s} (PEO) signal
intensity ratio for the PEO component will increase because
of the differences in electron mean free paths from these
levels, λ (C_{1s}) > λ (O_{1s}). Our results show that the C_{1s}
(PEO)/O_{1s} (PEO) signal intensity ratio remains invariant
with angle, θ. This result, alone, clearly indicates that
PEO is not covered by PS. If PEO is, therefore, exposed
at the air-polymer interface and the C_{1s}/O_{1s} ratio is not
angularly dependent, then the PEO in the copolymer must be
organized in a morphology such that its thickness is large
compared to the XPS sampling depth.

Let us now consider the two related models that are
shown in Figure 5. In these discontinuous models, both
components are exposed at the surface and they are
organized into domains. These domains are thick compared
to the XPS sampling depth and they occupy some fraction,
Δ, of the available surface area. These models result in
the predicted core level signal intensities shown in
Equations 4 and 5.[9]

$$I_o = f(\theta)[\Delta I_o^\infty (1-e^{-d/\lambda_o \cos\theta})] \qquad (4)$$

$$I_s = f(\theta)[(1-\Delta)I_s^\infty + \Delta I_s^\infty e^{-d/\lambda_s \cos\theta}] \quad (5)$$

Both models in Figure 5 are consistent with our observation
that the C_{1s} (PEO)/O_{1s} (PEO) signal intensity ratio for PEO
in the copolymer is invariant with angle θ. Tha major

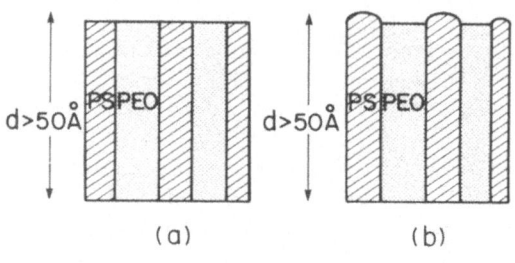

FIGURE 5

FIGURE 5

Models for the surface topography of polystyrene/polyethy-
lene oxide block copolymers.

difference between the models shown in Figures 5a and b is
that in the latter model, the PS domains are slightly
raised above the PEO domains at the surface, thus effectively
"shadowing" the PEO at high take-off angles, θ. This difference is significant because the model in Figure 5a predicts
no angular dependence in the PS and PEO signal intensities
whereas the model in Figure 5b predicts an increase in the
PS signal intensity at large values of θ. Data in Table 4
can be selected to support both of these models, but there
appears to be more evidence in support of the model shown
in Figure 5b.

Having established the general topography of the block
copolymer films, we can now use Equations 4 and 5 to calculate Δ and $(1-\Delta)$, the fractional surface areas occupied by
the PS and PEO components, respectively. The experimentally
measured intensity ratios for the PS (I_o/I_o^∞) and PEO (I_s/I_s^∞)
components in the copolymers were substituted into
Equations 4 and 5 and values of Δ and $(1-\Delta)$ were calculated
for the block copolymers cast from chloroform. The results
show a remarkable correlation between the calculated
fractional surface areas, Δ, and the molar composition of
the surface (mole % PS) determined by XPS, thus suggesting
that the fractional surface area occupied by the copolymer
components is proportional to the composition of the copolymer at the surface.[13]

CONCLUDING REMARKS

In this paper, we have briefly reviewed XPS techniques for characterizing polymer surfaces and for depth profiling compositional variations in multicomponent polymer systems in the top 50Å near the air-polymer interface. Our analysis of the PS-PDMS block copolymers suggested that the surfaces of these materials generally resembled pure PDMS and that the thickness of the PDMS surface overlayer was solvent dependent. More detailed XPS (θ) measurements on PS/PEO copolymers revealed that the surfaces were laterally inhomogeneous and that the components were isolated in thick domains whose fractional surface area was determined by the surface composition of the copolymers.

We are continuing to explore the application of XPS techniques to the characterization of multicomponent polymer systems and, in particular, to systems in which the components display appreciable mixing in the bulk.

ACKNOWLEDGMENT

We are indebted to Dr. D. Clark of the University of Durham and Dr. T. Davidson of the Xerox Corporation for many helpful discussions. We also appreciate the assistance of B. H. Fornalik in typing the manuscript.

REFERENCES

1.   K. Siegbahn, G. Nordlung, A. Fahlman, R. Nordberg, K. Hamrin, J. Hedman, G. Johansson, T. Berkman, S. E. Karlsson, I. Lingren and B. Lindberg, "ESCA, Atomic, Molecular and Solid State Structure Studied by Means of Electron Spectroscopy," Almquist and Wiksells, Uppsala, Sweden, 1967.

2.   D. T. Clark, Advances in Polymer Science, 24, 126 (1977).

3.   D. T. Clark, D. B. Adams, A. Dilks, J. Peeling and H. R. Thomas, J. Elect. Spect., 8, 51 (1976).

4.   D. T. Clark, A. Dilks, J. Peeling and H. R. Thomas, Faraday Disc., 60, 183 (1975).

5.   D. T. Clark and A. Dilks, J. Polym. Sci., Chem. Ed., 14, 533 (1976).

6.   D. T. Clark, J. Peeling and J. J. O'Malley, J. Polym. Sci., Chem. Ed., 14, 543 (1976).

7.   D. T. Clark, in "Advances in Polymer Friction and Wear," L. H. Lee, Ed., Plenum Press, New York, 1975.

8.   D. T. Clark, W. J. Feast, W. K. R. Musgrave and I. Ritchie, J. Polym. Sci., Chem. Ed., 13, 857 (1975).

9.   D. T. Clark and H. R. Thomas, J. Polym. Sci., Chem. Ed., 15, 2843 (1977).

10.   C. S. Fadley, R. J. Baird, W. Siekhaus, T. Novakov and S. A. Biegstrom, J. Electron Spectrosc. Relat. Phenom., 4, 93 (1974).

11.   J. J. O'Malley, R. G. Crystal and P. F. Erhardt, in "Block Polymers," S. L. Aggarwal, Ed., Plenum Press, N. Y., 1970, p. 163.

12.   H. R. Thomas and J. J. O'Malley, Amer. Chem. Soc. Div. Org. Coatings and Plastics Preprints, 38, 656 (1978).

13.   H. R. Thomas and J. J. O'Malley, Macromolecules, in press (1979).

PART III
PROPERTIES

CHANGING CONCEPTS OF CLUSTERING IN IONOMERS

A. Eisenberg

Department of Chemistry, McGill University

801 Sherbrooke St. W. Montreal, Que. Canada

I. INTRODUCTION

Since the appearance of ionomers over 10 years ago[1] considerable effort has been devoted to the elucidation of the structure of the ionic aggregates which are encountered in these materials. This is not surprising, in view of the fact that the incorporation of ions into organic polymers changes their properties profoundly. Over this time period, the accepted ideas concerning the structure of the aggregates have undergone several distinct changes, in spite of the fact that the detailed arrangement of the ionic groups within the aggregates has not, as yet, been elucidated, and neither has their precise size. These uncertainties have remained in spite of the fact that an extensive array of tools have been brought to bear on this problem. These tools include small angle X-ray scattering(SAXS), electron microscopy, IR, Raman and Mossbauer spectroscopy, a range of magnetic studies, and theoretical investigations. In addition, many of our ideas have come from an analysis of rheological and dielectric data.

In the first part of this presentation, a historical review will be given of the evolution of the concept of clustering in organic polymers. The second part will be devoted to a description of very recent research which gives some new insights into the problem of ion aggregation. Finally, some thermodynamic and kinetic aspects will be discussed.

Before concluding this introduction, it is worthwhile to introduce the definitions of two terms which will be used

231

extensively in the course of this discussion, i.e. multiplets
and clusters. Multiplets are considered to be small, tight
aggregates of ionic groups containing no organic material,
with the chains which are attached to the ions staying on
the outside of the aggregate. The multiplets act as
moderately strong crosslinks. By contrast, clusters are
taken to be large, more loosely interacting aggregates,
containing a relatively large amount of organic chain
material. Because of their size, these structures act not
only as crosslinks, but also as a strongly interacting
filler.

II. CLUSTERING IN A HISTORICAL CONTEXT

The polyphosphates were the first ion-containing
polymer system to have been investigated in some detail
in the bulk. The silicates had, of course, been studied
much earlier, but those are three-dimensional network, and
in most of the studies they were not treated in a manner
customary among polymer chemists. The phosphate studies
demonstrated the remarkable effect of ionic interactions
on the glass transition, e.g. an increase by ca 530°C in
going from $(HPO_3)_x$ to $(CA_{\frac{1}{2}}PO_3)_x$ while retaining the
identical backbone. Rheological studies of the simple
systems, e.g. $(NaPO_3)_x$, failed to reveal any anomalies other
than the rise in Tg. Thus, the problem of ion clustering
did not present itself in these materials; in the mid 1960's,
therefore, when this work began to appear, ion clustering
in polymers was not an issue. For the sake of the subsequent
discussion, it is worth recalling here that the dielectric
constant of the phosphate matrix is quite high, considerably
higher than that of the organic polymers to be discussed
subsequently. The studies of the polyphosphates in the
bulk from a polymer point of view were summarized in a
review of that period.[2]

By the late 1960's, several groups in both academic
and industrial laboratories were publishing studies on the
ethylene ionomers. Since these have been reviewed extensively in one article[3] and two books[4,5] no specific
references will be made to individual studies, only the
conclusions from the results will be described briefly, and
only insofar as they pertain to our understanding of the
structure of the clusters.

While one group of workers interpreted their results
to mean that the ionic groups existed as large clusters,
others believed that they were present only in small
multiplets. These conclusions were based on the appearance
of a new tan δ peak at ca 50-100°C with its position and
intensity increasing with ion content, on the stress relaxation results which showed that the systems behaved like
plastics containing substantial amounts of a reinforcing
filler, and, above all, on SAXS and electron microscopy
data. Many additional studies were performed, ranging all
the way from water absorption to spectroscopy, with the
results being interpreted in the same way. Interestingly
enough, both groups of workers used essentially the same
results to advance their viewpoint. The controversy
continued until the early 1970's, when the consensus shifted
gradually in favor of the large cluster advocates. In the
early 1970's, several publications appeared dealing with the
styrene ionomers.[6] Work on these systems is still continuing, but even the early results showed that two different
types of aggregates had to be present in these materials.
This conclusion was based primarily on discontinuities in
the rheological behavior at ca 5-6 mol% of the ionic
comonomer. For example, in stress relaxation below 6% the
ions act as temporary crosslinks, with the material still
subject to time-temperature superposition, while above that
concentration the polymers are thermorheologically complex
with the ions acting as a strong reinforcing filler.
Furthermore, the W.L.F. constants, when plotted as a function
of ion concentration, show a pronounced discontinuity at that
ion content as do melt rheology studies, specifically in a
plot of the temperature by which the ionomer samples have
to be heated over and above the equivalent ester samples
(ΔT_η) to get superposition in the G' vs ω behavior. A
sharp upswing in the ΔT_η values is observed at ca 6%. SAXS
reveals the presence of a low angle peak only above ca 5%,
and both the glass transition and the density show appreciable changes in slope at that concentration, as can be
seen in Fig. 1. Finally, water uptake studies at room
temperature show an absorption of only one water molecule
per ion pair below 6%, while above that concentration, in
the region of 6-10%, that number increases to the range of
3-5.

The above mentioned experiments have been interpreted
initially as suggesting that below an ionic comonomer content

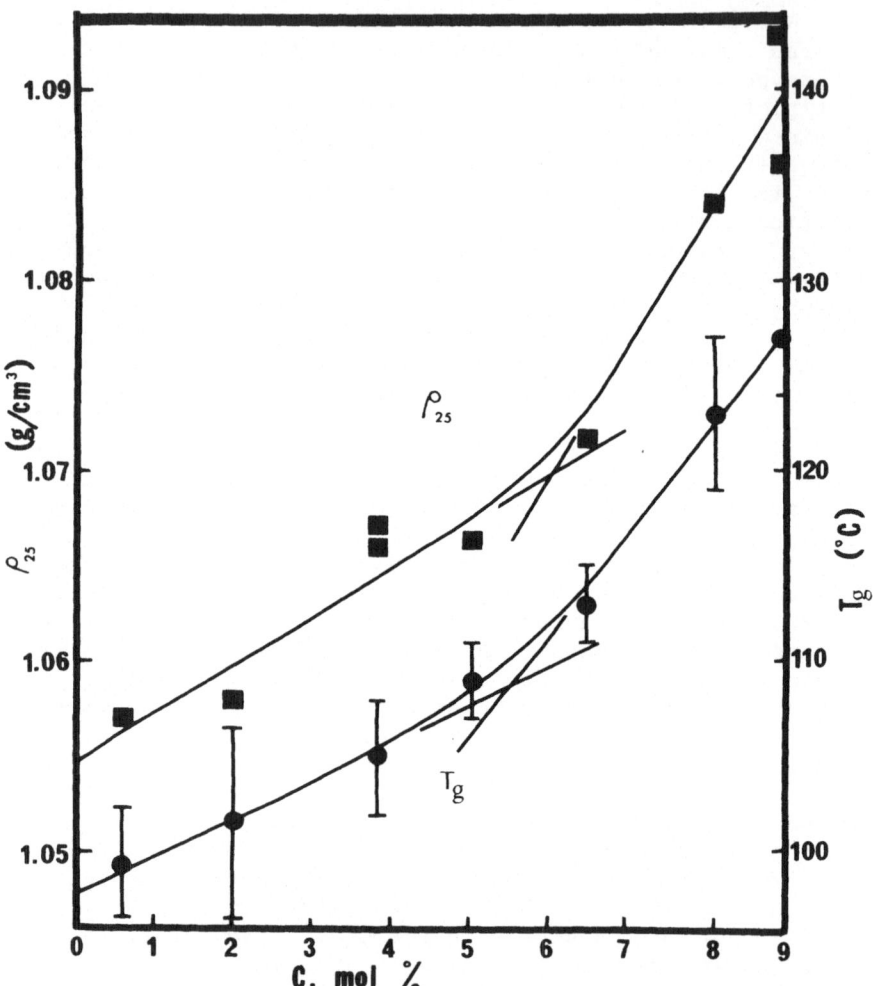

Fig. 1 Glass transition and density vs. ionic comonomer content for styrene ionomers.

of 6 mol% the ions are aggregated as small tight multiplet, whereas above that concentration they are present in large clusters. Furthermore, preliminary results on the ionized ethyl acrylate-acrylic acid copolymer system suggested that the transition region in that polymer was located in the range of 10-15 mol% of ions.[7] Thus, a picture emerged

which implied that the critical concentration above which
the ions give large clusters varies with the dielectric
constant, from <1% for ethylene, through 6% for styrene,
10-15% for ethyl acrylate all the way to the unclustered
polyphosphate system. A simple theoretical treatment[8] of
ion clustering was in qualitative agreement with the above
findings, especially in terms of the existence of a critical
concentration below which clusters would not be fevored.

III. THE PRESENT VIEW OF CLUSTERING

The first semi-quantitative idea of the distribution
of ions among clusters and multiplets emerged from a
dielectric study of the styrene ionomers.[9] In that work,
it was shown that two dielectric peaks were present in the
glass transition region of the materials. Both peaks were
present above ca 2 mol% of ions. The peak maxima at constant
frequency rose with increasing ion content, the rate of
increase of the high-temperature peak (T_h) being somewhat
faster than that of the low-temperature one (T_ℓ). The
integrated intensity of the T_ℓ peak rose up to an ion
content of ca 48%, after which it levelled off, while that
of the T_h peak rose at an accelerated rate. Most significantly, the sum of the integrated intensities of the two
peaks correlated very well with the total ion concentration,
the correlation coefficient being 0.975. It was, thus,
very tempting to conclude that the T_ℓ peak was due to the
ions in the multiplets, while the other originated from
the clusters. The reason for that assignment was based
primarily on the consideration that the multiplets are
dispersed in the organic matrix, where they act as corss-
links, and the matrix softens at a lower temperature than
the clustered regions. Thus, assuming that the dielectric
response per ion pair is the same in the cluster and the
multiplet, and depending on how one substracts the tan δ
peak at the glass transition of pure PS, one gets a range
of values for the number of ions in the clusters and in
the multiplets as a function of ion content. This is shown
in Figure 2.

The assumption that the dielectric response per ion
pair is the same in the clusters and the multiplets is,
of course, subject to question. It is therefore, most
gratifying that the results of a completely different
experiment, i.e. Raman spectroscopy, yield identical results

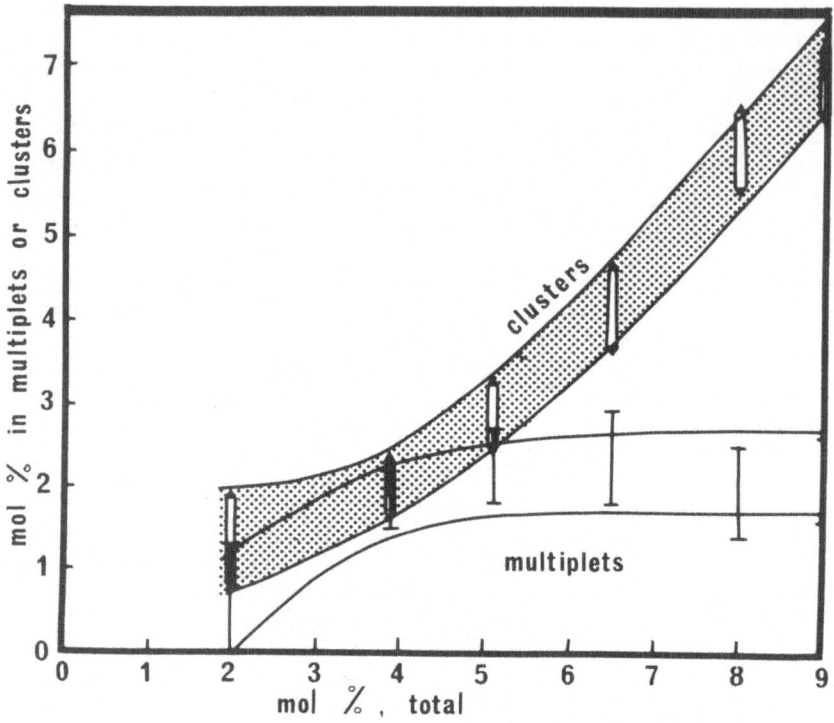

Fig. 2 Mol % ions in multiplets or clusters vs total
mol % of ionic comonomer in the styrene ionomers (from
dielectric data).

in terms of the concentration of ions in multiplets and
clusters.[10] Very recently it was found that the presence
of ions in the styrene ionomers gives rise to two bands,
located at 166 cm^{-1} and 254 cm^{-1}. These have been ascribed
to the clusters and the multiplets, respectively. When the
integrated intensities of these bands are normalized with
respect to a band due to the benzene ring, and the relative
intensities plotted against the total ion content, a plot
very similar to that given in Fig.2 is obtained. Furthermore, the sum of the integrated intensities shows a very
good correlation with the total ion content ($\gamma \approx 0.99$).
Finally, the intensity of the "cluster mode", when plotted
against the integrated intensity of the high temperature
dielectric tan δ peak, gives a straight line with a

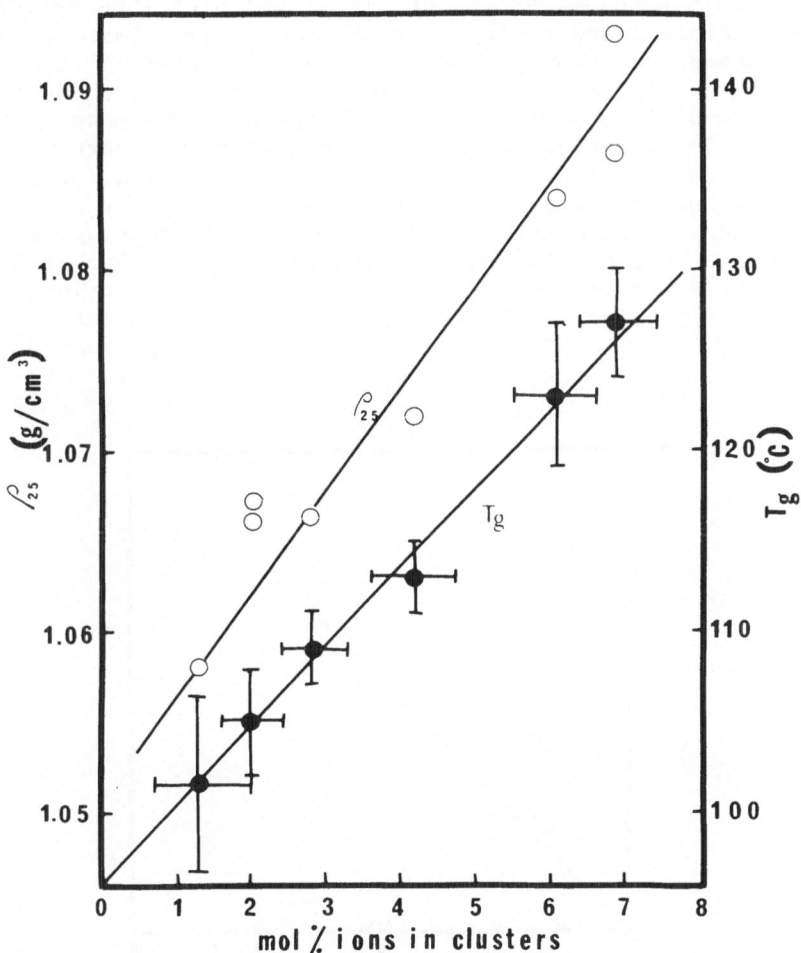

Fig. 3 Glass transition and density vs mol % ions in clusters for the styrene ionomers.

correlation coefficient γ>.999. The agreement of these
two different experiments is most gratifying, and supports
the proposed assignments.

At this point, since the distribution of ions between
the multiplets and the clusters is known approximately, one
can attempt correlations between the various properties
which are affected by the presence of ions and the content
of the ions in the clusters or multiplets. The most
instructive correlations are obtained with the content of
ions in the clusters. Fig. 3 shows this relation for the
density and the glass transition, while Fig. 4 gives it for
the temperature shift obtained from the melt rheology studies
discussed before. All three plots are linear, suggesting
that it is the clusters that exert the greatest influence
on the properties of the ionomers, not the multiplets.

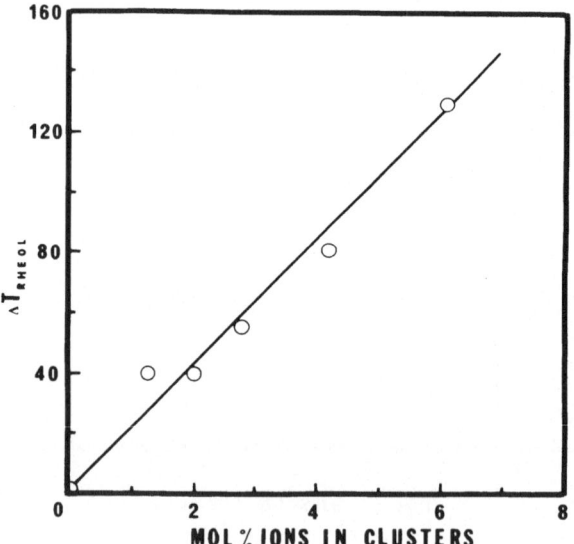

Fig. 4 ΔT_η vs mol % ions in clusters for the styrene ionomers.

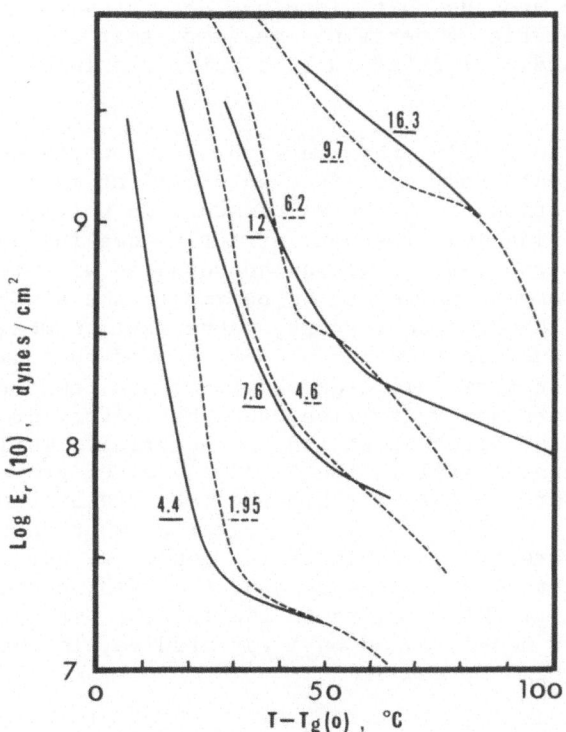

Fig. 5 Young's Modulus (10 sec) vs $T-T_g$ for the styrene ionomers (dotted lines) and the ethyl acrylate ionomers (solid lines) for various ion contents.

In regard to the effect of ions in media of varying
dielectric constant, a comparison of the modulus temperature plots of two non-crystalline ionomer systems is perhaps
most instructive. Fig. 5 shows this for various ion concentrations in the styrene (S) and ethyl acrylate (EA)
ionomer system.[7] It is clear that the ion content in the
EA copolymer has to be twice as high as in the S case to
obtain a similar effect. The distribution of ions between
the clusters and the multiplets has not, as yet, been
explored, but Fig. 5 certainly suggests that the ratios of
the two should be related in both materials in very similar
ways.

Before concluding this section, it is worth noting
that one recent study has given an excellent idea of the
size distribution of the metal clusters in polymers.[11]
The material involved was not, strictly speaking, an ionomer,
because the iron (as $FeCl_3$) was coordinated with vinyl
pyridine comonomer units on the polymer. The study, which
utilized Mossbauer spectroscopy, magnetization measurements,
SAXS, SANS, and electron microscopy, showed that the iron
was present in three different environments, that the
"clusters" were noncrystalline, and that a very broad
distribution of sizes exists in the material. While most
of the aggregates fall in the 50–100 Å range, progressively
smaller numbers of larger aggregates are present, ranging
in size up to 1500 Å. It is not clear to what extent these
conclusions are transferable to the ionomers, but are certainly suggestive of what should be expected in these
materials also. With regard to the latter, perhaps the most
sophisticated model, based on X-ray studies, is that of
MacKnight, Taggart, and Stein.[12]

IV. KINETIC AND THERMODYNAMIC ASPECTS OF CLUSTERING

A recent investigation gives some insights into one
aspect of the kinetics of cluster formation, both above
and below Tg. In a study of the expansion coefficients of
the styrene ionomers as a function of ion content[13] it
was found that the equilibrium values were independent of
ion content. However, whenever the sample had been allowed
to remain at room temperature for a few hours after cooling
from above Tg, and subsequently heated again above the glass
transition temperature, the first run yielded a value of
the liquid expansion coefficient α_ℓ ca 25% lower than that

obtained on subsequent cycling above room temperature. The
sample had to be heated to ca Tg + 60 and then cooled before
the equilibrium value was attained. Storage at Tg − 30 for
even five hours failed to yield the low value of α_ℓ; the
sample had to be stored at room temperature for a few hours.
By the same token, below Tg + 60 no change in the expansion
coefficient was observed even on the first run. It is
tempting to conclude from these findings that those clusters
which influence the expansion process in the styrene ionomers
are quite stable up to ca Tg + 60, but that once they fall
apart it is necessary to keep the sample well below Tg, in
this case at room temperature, before the clusters are
reformed. It should be borne in mind, however, that a wide
distribution of cluster sizes exists, most probably, and
that not all of these will affect all properties equally.
Thus, clusters in one range of sizes might affect the
expansion coefficient, while another range of sizes affects
the rheological behavior, etc.

The variation of the WLF constants with temperature
presents valuable insight into the thermodynamic aspects of
clustering. An early study of the mechanical properties (6)
revealed that the constants C_1 and C_2 change only slightly
with increasing salt content of the comonomer up to ca 5-6
mol% of ions, that a discontinuity occurs in that region,
and that the values of both constants increase considerably
in the range of 6-9 mol%. At that time, the phenomenon was
ascribed to the onset of ion clustering at ca 6 mol%. A
more recent dielectric investigation, which showed that
clusters and multiplets coexist, postulated that the clusters
decomposed progressively at elevated temperatures into
multiplets, and that as a result of this overall change in
the structure the effective glass transition varied with the
temperature in a manner given by the equation

$$Tg = Tg° + a(T-Tg°)$$

where Tg and Tg° are the effective and the measured glass
transition, respectively. By inserting this relation into
the modified WLF equation for the activation energy and
comparing the result with the original equation for the
activation energy, one can obtain a relation between the
observed C_1 for the ionomer, C_1^{obs} and the C_1 for pure PS,
C_1^{PS}, given by the equation

$$C_1^{obs} = C_1^{PS}/(1-a)$$

with the same relation holding for C_2. It should be recalled that Tg° is the measured glass transition of the ionomer, i.e. in effect the position of the low temperature tan δ peak. The values of "a" calculated that way range from ca 0.3 for the 1.9 mol% sample to ca 0.8 for that containing 9.7 mol% salt.

Confirmation for the above view comes from a very recent and more extensive study of the dielectric properties as a function of degree of neutralization.[14] If that work, it was shown that even in the neutralized samples, the peak frequencies for the low temperature peak, when plotted against temperature, gave a curve which could be described by the WLF equation with "normal" parameters, i.e. close to those of pure polystyrene, even above 6 mol% of ions. It should be recalled that the low temperature peak occurs at ca 130-160°C (at 100 Hz). Furthermore, an inspection of the plots of log a_t vs T-Tg shows that up to Tg + 40°C, the curves for all the samples, independent of ion content, fall on practically the same line, and only above that temperature do major differences appear between the samples below 6 mol% and those above that ion content.

V. CONCLUDING COMMENTS

Some comments concerning recent developments in our view of cluster morphologies are in order. In an early study, Stein et al have shown that stretching of an ionomer sample does not result in an orientation of the clusters. This suggested that the clusters are, most likely, spherical, a view supported by the recent study of Pineri and Meyer for the chelated polymer. A more recent SAXS study, however,[15] showed clearly that orientation effects are encountered. This finding is consistent either with a non-spherical structure of the clusters, or with a sphere which deforms when the sample is stretched. It is clear that much further work remains to be done on the elucidation of the shape of the clusters, as well as the geometrical arrangements of the components, i.e. the ions and the polymer chains. This is true not only of the "classical" ionomers, i.e. those based on ethylene, styrene, or the rubbers, but also of the newly developed materials which contain ionic domain plasti- cizers, consisting of materials such as EPDM ionomers plasti- cized with zinc stearate.[16] The field should remain a most challenging one in the foreseeable future.

REFERENCES

1. R.W. Rees and D.J. Vaughan, A.C.S. Polymer Preprints, 6 (1), 287 (1965).
2. A. Eisenberg, Adv. Poly. Sci., 5, 59 (1967).
3. E.P. Otocka, Macromol. Sci.-Rev. Macromol. Chem., C5 (2), 275 (1971).
4. L. Holliday, Ed. "Ionic Polymers" Applied Science Publisher, 1975.
5. A. Eisenberg and M. King, "Ion-Containing Polymers", Acad. Press, 1977.
6. Ref. 5, P. 414-162.
7. A. Eisenberg, H. Matsuura, and T. Tsutsui, to be published.
8. A. Eisenberg, Macromolecules, 3, 147 (1970).
9. I.M. Hodge and A. Eisenberg, Macromolecules, 11, 289 (1978).
10. A. Neppel, I.S. Butler, and A. Eisenberg, J. Poly. Sci., In Press.
11. C.T. Meyer and M. Pineri, J. Poly. Sci., Polym. Phys. Ed., 16, 569 (1978).
12. W.J. MacKnight, W.P. TAggart, and R.S. Stein, J. Poly. Sci. Polymer Symposia, 45, 113 (1974).
13. A. Eisenberg and E. Trepman, J. Poly. Sci. Polym. Phys. Ed., 16, 1381 (1978).
14. K. Arai and A. Eisenberg, to be published.
15. E.J. Roche, T.P. Russell, R.S. Stein, and W.J. MacKnight, Polymer Preprints, 19 (2), (1978).
16. R.D. Lundberg, H.S. Makowsky et al, Polymer Preprints, 19 (2).

ACKNOWLEDGEMENT

It is a pleasure to acknowledge that this work was supported by the Army Research Office.

MICRODOMAIN STRUCTURE AND SOME RELATED PROPERTIES OF BLOCK COPOLYMERS

Hiromichi Kawai and Takeji Hashimoto

Department of Polymer Chemistry,

Kyoto University, Kyoto 606, Japan

INTRODUCTION

Block copolymers composed of incompatible components of A and B form in general a "pseudo two-phase" structure in solid state as a consequence of microphase separation in solidification process. There exists interfacial domain boundary region of a certain thickness where the incompatible components are mixed in between the regions composed of pure A and B segments. Origin of such interphase has recently been studied intensively on the basis of statistical thermodynamics by Meier(1), Helfand(2), and others. Number of works on the mechanical properties of block and graft copolymers have also proposed the existence of the interphase(3-9).

In this article, the analysis of microdomain and domain-boundary structures of amorphous block copolymers is first investigated by means of small angle X-ray scattering and electron microscopy. The structures, especially for alternating lamellar microdomains of a series of A-B type block copolymers of styrene and isoprene, are discussed in comparison with those predicted from the current theories of domain formation by Meier and Helfand.

Second, the strain-softening phenomenon of a styrene-butadiene-styrene tri-block copolymer and its blend with polystyrene both having alternating lamellar microdomains of the two components; i.e., "strain-induced plastic-to-rubber transition" is investigat-d in terms of change of the alternating lamellar microdomains to fragmented sty-

rene domains dispersed in a butadiene matrix. An inverse
transition from rubber to plastic, which is associated
with "healing effect" of the fragmented structure after
releasing from high %-elongations even at temperature as
low as 60 $^\circ$C for duration as short as 10 min, is rationalized in light of domain-boundary relaxation mechanism activated with anomalous increase in volume fraction of the
interphase region in association with the fragmentation of
the styrene lamellar domains. An application of the "plastic-to-rubber" and "its inverse" transitions to the development of new material is demonstrated for a rubber
toughened plastics.

INVESTIGATION OF LAMELLAR SPACING AND INTERFACIAL THICKNESS
OF STYRENE-ISOPRENE BLOCK COPOLYMER FILMS CAST FROM SOLU-
TIONS AND ITS COMPARISON WITH CURENT THEORIES

As has been discussed previously(10-13), the X-ray
scattering from assemblies of particles may be given by

$$I(\underline{s}) = N[<f(\underline{s})^2> - <f(\underline{s})>^2]$$
$$+ (1/v)<f(\underline{s})>^2 Z(\underline{s}) * |S(\underline{s})|^2 \qquad (1)$$

The first term of Eq.(1), a diffuse back-ground scattering
is generally much smaller and less angularly dependent than
the second term, and $S(\underline{s})$, the shape amplitude related to
the size and shape of the assembly of the particles varies
with scattering angle s much more rapidly than $Z(\underline{s})$, the
paracrystalline lattice factor, so that

$$I(\underline{s}) \cong (1/v)<f(\underline{s})>^2 Z(\underline{s}) * |S(\underline{s})|^2 \cong N<f(\underline{s})>^2 Z(\underline{s}) \quad (2)$$

For a scattering at large angle from disordered system,
$Z(\underline{s})$ tends to reduce to unity, and it follows

$$I(\underline{s}) \cong N<f(\underline{s})>^2 \qquad (3)$$

Eq.(3) is the so-called independent scattering approximation relating the scattering intensity $I(\underline{s})$ and number of
the particles per radiated volume N to the scattering amp-
litude of individual particle $f(\underline{s})$. $f(\underline{s})$ can be given in
terms of Fourier transform of electron density fluctuation
of the particle along $\underline{x}$ as follows:

$$f(\underline{s}) = \mathcal{F}[\eta(\underline{x})] = \int_V \eta(\underline{x})\exp[2\pi i(\underline{s}\cdot\underline{x})]d\underline{x} \qquad (4)$$

The electron density fluctuation $\eta(\underline{x})$ may be given by a convolution product of an electron density function $g(\underline{x})$ of a particle having no interphase with a smoothing function $h(\underline{x})$ associated with density transition at the interphase. Assuming $g(\underline{x})$ and $h(\underline{x})$ to be a box type function with height of ρ_o and width of 2a and a Gaussian type function with standard deviation σ, then $\eta(\underline{x})$ can be given by a sigmoidal function as follows:

$$\eta(\underline{x}) = (\rho_o/2) - (\rho_o/\pi^{1/2})\ \mathrm{Erf}[(a+x)/(2^{1/2}\sigma)] \qquad (5)$$

In order to compare the domain-boundary structure defined here in terms of the sigmoidal function with those proposed by Meier and Helfand, let one further define the interfacial thickness Δt as follows:

$$\Delta t = \rho_o/[d\eta(x)/dx]_{x=a} \qquad (6)$$

Fig. 1 shows the comparison of the three kinds of domain-boundary structures with each others under a paricular condition so as to make Δt identical and ρ_o as unity.

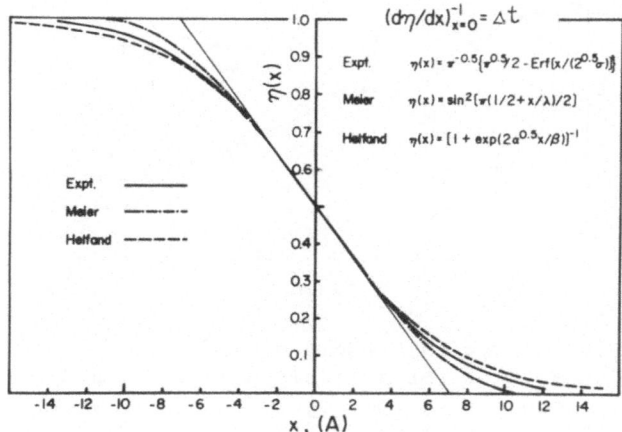

Fig. 1. Comparison of three kinds of domain-boundary structures

The scattering intensity distribution $I(\underline{s})$ from alternating lamellar microdomains at relatively large scattering angles is given for randomly or highly oriented system as

$$I(\underline{s}) = (\text{const}) s^{-4} \exp(-4\pi^2 \sigma^2 s^2) \qquad (7)$$

or

$$I(\underline{s}) = (\text{const}) s^{-2} \exp(-4\pi^2 \sigma^2 s^2) \qquad (8)$$

Eq. (7) or (8) indicates that $\ln[I(\underline{s})s^4]$ vs s^2 or $\ln[I(\underline{s})s^2]$ vs s^2 plot gives a straight line relation, and that σ can be determined from the slope of the straight line. Further Δt can be obtained from σ by a following relation:

$$\Delta t = \sqrt{2\pi}\sigma \qquad (9)$$

Fig. 2 shows a typical result of the $\ln[I(\underline{s})s^2]$ vs s^2 plot for a highly oriented system after corrections of slit smearing and back-ground scattering of observed SAXS intensity distribution.

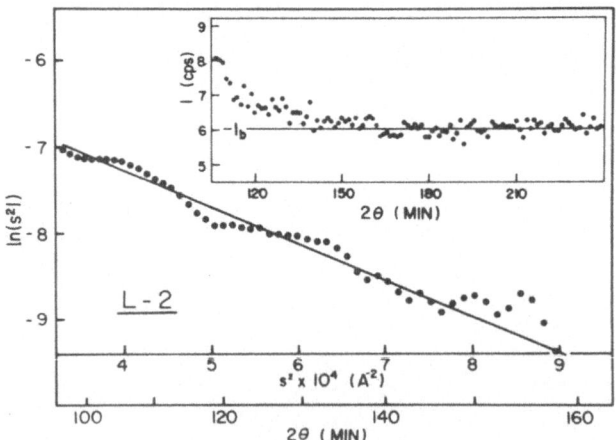

Fig. 2. Typical result of $\ln[I(\underline{s})s^2]$ vs s^2 linear plot for a highly oriented lamellar microdomain system. Upper figure shows an elimination of back-ground scattering

Figs. 3 and 4 show comparison of the domain properties, the lamellar spacing D and the interfacial thickness Δt,

with those predicted from the theories for a series of A–B
type block copolymers of styrene and isoprene cast from
toluene solutions. All of the block copolymers have fractional composition of the two components near 50/50 but
differ in total molecular weight in a range from 2 to 10 x
10^4. As can be seen in Fig. 3, the lamellar spacing D is
found to agree well with theoretical ones in its molecular
weight dependence, though the spacing itself is found to be
a little less than the theoretical values. In detail, the
spacing is found to follow 0.64th power of total molecular
weight, while theoretical ones are 0.59 and 0.66 poweres
for Meier's and Helfand's theories, respectively. Small
discrepancy between the observed and calculated spacings
must be ascribed to the difficulty in choosing proper values
of molecular and thermodynamic parameters, such as Flory-
Huggins interaction parameter, χ, for theoretical calculation for this particular system.

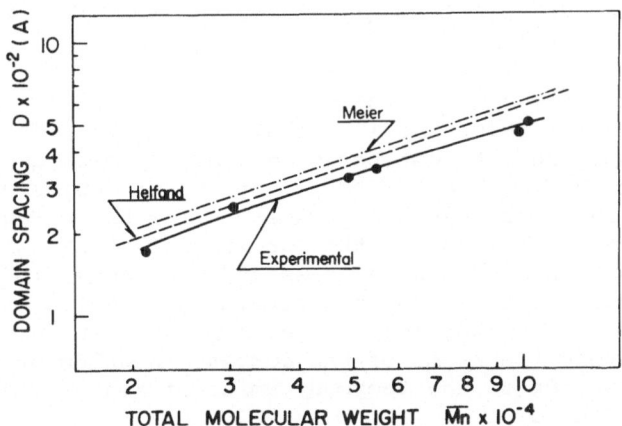

Fig. 3. Comparison of observed lamellar
spacing with calculated ones

On the other hand, the interfacial thickness Δt is
found, as can be seen in Fig. 4, to be a constant of around
25 A over the range of total molecular weights. The thickness is considerably larger than that predicted from the
Helfand's theory as a constant of around 15 A over the molecular weight range, but is smaller than that predicted from
the Meier's theory as decreasing from 50 to 30 A with increasing molecular weight within the range. Although the

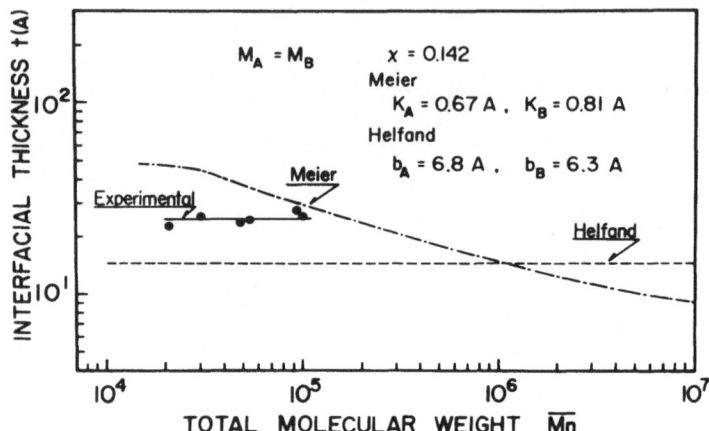

Fig. 4. Comparison of observed interfacial
thickness with calculated ones

discrepancy between the observed and calculated results for
the interfacial thickness is rather obvious than for the
lamellar spacing, it must be noted that the experimental
difficulty, besides the difficuly of proper choice of the
molecular and thermodynamic parameters for the calculation,
is also rather serious for the interfacial thickness than
for the lamellar spacing. At present, therefore, it may be
concluded that the microdomain structures, including the
domain-boundary structure, can be explained, at least semi
quantitatively, in terms of the current theories of domain
formation basing on the thermodynamics of random fleight
chain statistics.

For the above calculations of D and Δt, the Flory-
Huggins interaction parameter χ and the Kuhn's statistical
segment length b_i are assumed to be 0.142 and 6.8 A and
6.3 A for PS and PI, respectively. The values are, however,
based on literature ones for cis-1,4 polyisoprene and are
not necessarily sufficient for the present copolymers synthesized by anionic polymerization in a polar medium of
THF to contain high 1,2 and 3,4 microstructures. Experimental evaluations of the parameters for this particular
series of block copolymers are now carrying out in this
laboratory.

STRAIN-INDUCED PLASTIC-TO-RUBBER TRANSITION OF AN SBS TRI
BLOCK COPOLYMER AND ITS INVERSE TRANSITION AFTER RELEASING
     It has been shown that some block copolymers and their
blends with corresponding homopolymers exhibit the so-
called strain-softening effect(14-17). Moreover, the strain
softened specimens exhibit a healing effect, in that properties of the original undeformed specimens are recovered
upon removal of the applied stress(14,16,17). The effect
has been postulated as the reformation of the original
microdomain structure. Here, the strain-softening and,
especially, the healing processes are investigated by means
of electron microscopy and small angle X-ray scattering. The
healing effect is rationalized in light of domain-boundary
relaxation mechanism(18,19).

$\leftarrow$ 0.5µ $\rightarrow$

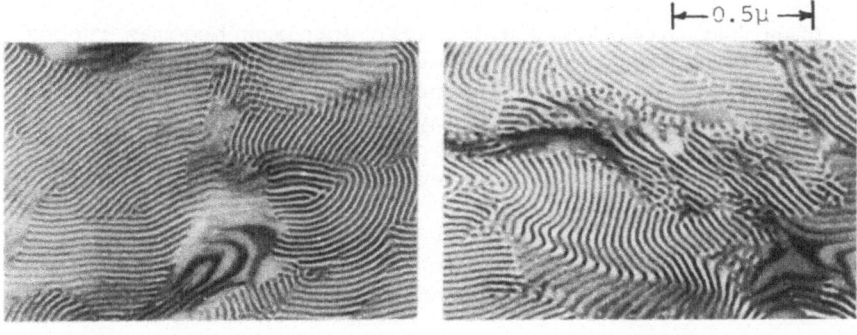

Fig. 5   Fig. 6

     Figs. 5 and 6 show electron micrographs of ultrathin
sections of film specimens of an SBS block copolymer and its
blend with PS. The average molecular weights of the SBS and
PS are 14-30-14 and 20.4 in units of thousands. The film
specimens were spin-cast from 10% solutions of SBS and its
blend with 20% PS in a mixed solvent of tetrahydrofuran and
methyl ethyl keton (90/10 by volume). As can be seen in the
figures, both specimens have alternating lamellar domains
of the two components, oriented randomly. The styrene lamellae of the SBS-PS blend specimen are found to be thicker
than those of the SBS specimen, due to solubilization of the
PS into styrene lamellar domains of the SBS.

     Fig. 7 shows the cyclic tensile stress-strain behavior
of the SBS and SBS-PS specimens measured in the first and
second stretch cycles. The experiment was conducted at room

temperature at a strain rate of 50%/min. For clarity only
the stretching half-cycles are shown. As can be seen in the
first stretching half-cycle in the figure, after the yielding has taken place there is a plateau region followed by
a rapid increase of the stress. The block copolymer exhibits
a yield point at around 5% strain, after which the stress
decreases until about 20% strain. The stress then remains
constant with further elongation up to about 180% strain.
This is typical plastic-like behavior. When the applied
stress exceeds the yield stress, necking suddenly appears
at a localized region in the specimen, which subsequently
grows continuously until the whole specimen is covered. The
necking process gives rise to the plateau region observed in
the stress-strain curve. Upon further stretching, the stress
rises rapidly and fracture soon follows. If the deformation
process is reversed, there is now substantial strain recovery. Thus the initially plastic-like specimen becomes rubber
like at the completion of the necking process. The stress-
strain behavior during the second cyclic deformation is also
rubber-like and does not show any yielding and necking phenomena except for a small but rapid increase in stress at
the beginning of the second stretching half-cycle.

The blend specimen exhibits similar strain-induced
plastic-to-rubber transition, as clearly seen in the figure.
Due to the higher overall PS content of the specimen, it

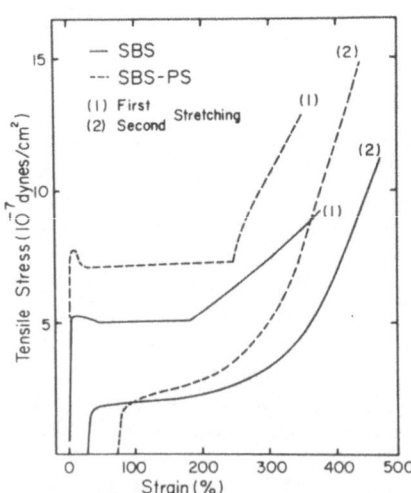

Fig. 7. Cyclic tensile stress-strain behavior

shows a higher value of Young's modulus (initial modulus),
yield stress, stress at the plateau region, and the tangent
modulus after completion of the necking. The observed residual strain after removal of the applied stress is also
higher than the SBS specimen.

|← 0.5μ →|

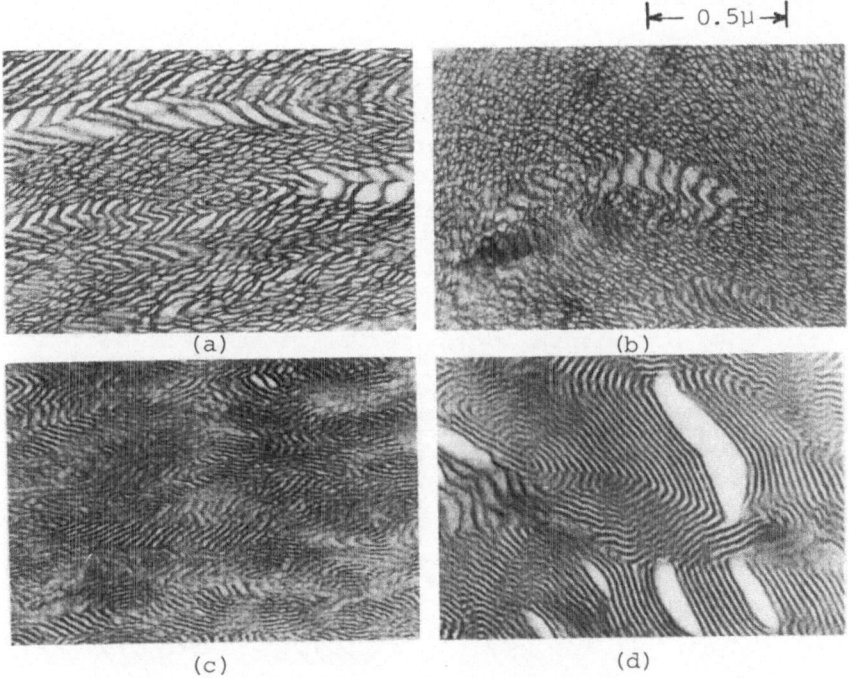

Fig. 8. Electron micrographs of SBS specimen; (a) stretched
to 85%, (b) stretched to 500%, (c) released from 600%, and
(d) released and annealed at 100 $^{\circ}$C for 2 hrs. The stretch
direction is horizontal. Ultrathin sectioning was made parallel to the stretching direction and normal to the film
surfaces.

Stretching the SBS specimen by 85% elongation produces
irregular deformation of the lamellar microdomains accompanied by shearing, kinking, disruption and orientation of the
microdomains, as seen in Fig. 8(a). These deformation processes are responsible for the yielding and necking of the
specimen and are dominant until the necking is completed.
Upon further stretching, fragmentation of the lamellar micro
domains prevails. Finally fragmented polystyrene domains are

dispersed in a matrix of polybutadiene, as can be seen in
Fig. 8(b). These fragmetation processes must have been responsible for the plastic-to-rubber transition. The polystyrene frgments act as surface-active filler particles for
polybutadiene, and the elasticity of the specimen is essentially entropic in origin.

It is also observed in Fig. 8(c) that the fragmented
polystyrene domains are transformed into the original lamellar domains after "resting" in the unstretched state for
several days at room temperature. The reformed structure is
more perfect when it is annealed at elevated temperature,
as seen in Fig. 8(d), and less perfect for the SBS-PS. The
reappearance of lamellar morphology, however, is obvious in
both types of healing process.

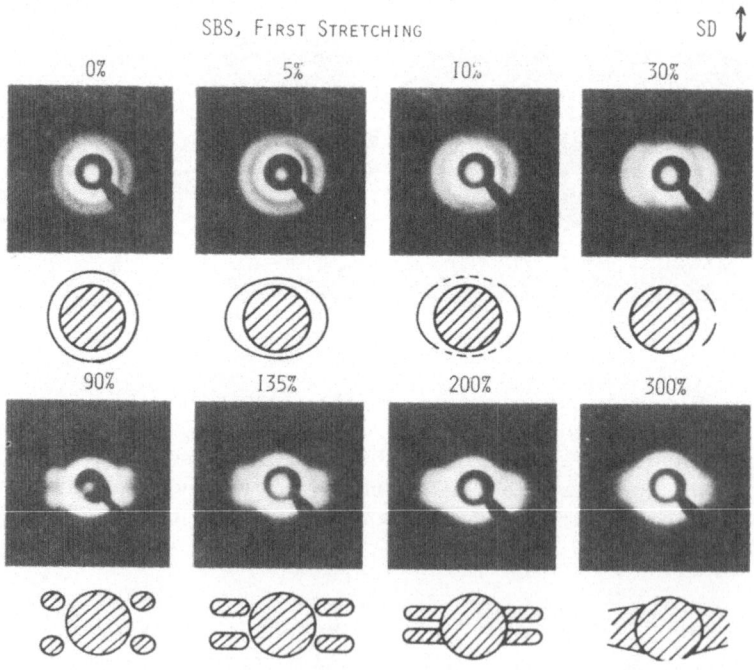

Fig. 9. Change of the SAXS patterns of SBS during the
course of the first stretching cycle. The stretching
direction is vertical.

Fig. 9 shows the change in SAXS patterns upon stretching the SBS specimen in the first stretch half-cycle. The figure also includes schematic representations of the SAXS patterns with only the first-order scattering maximum. The numbers in the figure indicate percent-elongations of the specimen. Upon stretching the meridional SAXS maxima tend to shift toward smaller angles. Its intensity decreases until 30% elongation where it disappears. This indicates that the lamellae, originally oriented with their boundaries normal to the stretching direction, increase their spacing with elongation. However, the spacing also becomes irregular due to such deformation processes as shearing, kinking, and destruction of the lamellae. On the other hand, the equatorial SAXS maxima remain nearly invariant under the same conditions, which indicates that the spacings do not vary for those lamellae which were originally oriented parallel to the stretch direction. This accounts for the volume dilatation of the specimen.

At higher elongations, the equatorial scattering maxima disappear due to the deformation processes discussed above. The scattering patterns in the plateau region of the stress strain curve after the yield point are characteristic of lamellar surfaces inclined to the stretch direction. With increasing %-elongation the scattering lobes tend to be elongated parallel to the equator, i.e., the lateral breadth of the lobe increases, indicating that the lamellae are fragmented so that their lateral continuity decreases, as quantitatively discussed elsewhere(19). Upon further stretching scattering lobes disappear. The scattering pattern becomes diffuse, and is more or less independent of the azimuthal angle. The implication is that fragmented polystyrene domains are randomly dispersed in the rubber matrix. This is consistent with the conclusion obtained by a qualitative examination of the electron micrographs in Figs. 8(a) and 8(b). The polystyrene domains now act as filler particles, on which surfaces the polybutadiene chains (the middle block segments of the SBS) must be anchored and the specimen exhibits rubber-like elasticity.

Fig. 10 summarizes schematically the structure changes occuring in the strain-induced plastic-to-rubber transition. The change in structure from (a) to (b) illustrates the initial stage of deformation of the microdomain up to the yield point, as observed in the expansion of the lamellar spacing

for the lamellae oriented perpendicular to the stretch direction. The changes in structure from (b) to (d) illustrate
the expected changes in the yielding and necking processes
in which the irregular deformation of the lamellar microdomains occurs. The deformation involves kinking, shearing,
destruction, and orientation of the lamellae. Finally, fragmented polystyrene domains are randomly dispersed in the
rubber matrix, as illustrated in (e), and act as surface-
active filler particles for the polybutadiene chains in the
matrix.

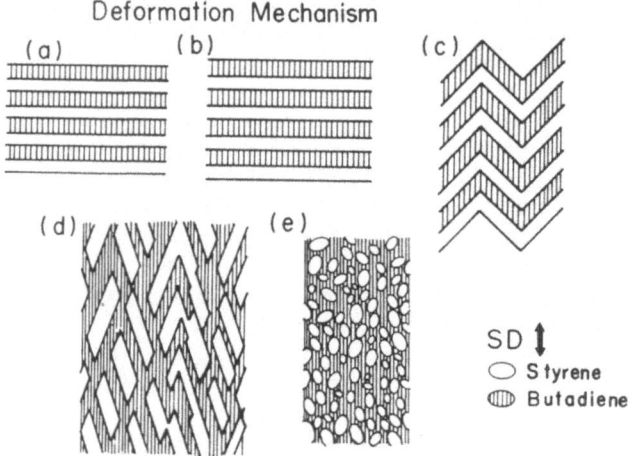

Fig. 10. Schematic representation of the deformation processes involved in the strain-induced
plastic-to-rubber transition.

        To further examine the effect of healing on morphology,
we show in Figs. 11(a) and 11(b) the SAXS patterns of the
SBS and SBS-PS specimens released from 355% elongation and
rested at room temperature for a few days, and in Figs. 11
(c) and 11(d) the patterns of the SBS and SBS-PS specimens
released from 390%- and 500%-elongations, respectively, then
both annealed at 59 °C for 5 hr. In all figures the stretch
direction is vertical. As can be seen in Figs. 11(a) and 11
(b), no significant change in the SAXS patterns is observed
upon healing, except for a faint reappearance of the first
order maximum at equatorial zone for the SBS specimen. There
fore, the structure does not become sufficiently regular to

give the scattering maxima higher than the first-order. As
expected the reformation process is slow at room temperature. On the other hand, Fig. 11(c) shows that the annealing
at 59 $^{\circ}$C substantially restores the original lamellar domain
structure, especially for the SBS specimen, though the lamellar spacing parallel to the stretch direction is still
slightly expanded than that normal to it. It is of interest
to note that structural reformation does occur at a temperature as low as 59 $^{\circ}$C, which is considerably lower than the
glass-transition temperature of polystyrene.

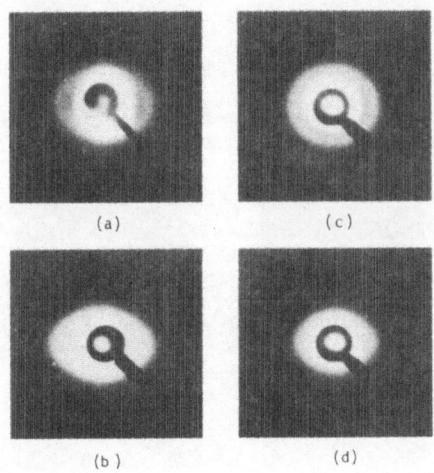

Fig. 11. SAXS patterns of SBS and SBS-PS taken
during the healing process at room temperature
and 59 $^{\circ}$C.

Fig. 12 shows electron micrographs of ultrathin sections of the SBS-PS specimens, displaying the healing effect
after releasing from 500% elongation (the stretch direction
is horizontal). Here (a) is after healing at room temperature for a few days and (b), (c) and (d) show the healing
effect followed by annealing at 60, 88, and 116 $^{\circ}$C, respectively, all for 5 hr. Although the initial stretching of the
specimens up to 500% elongation is larger than the SAXS test
in Fig. 11, the structural reformation is generally much
more pronounced for the SBS specimen than for the SBS-PS
specimen, and its general trend is consistent with that

expected from SAXS patterns shown in Fig. 11. For example,
in Fig. 12(a) the reformed domain structure is still lar-
gely oriented in the stretch direction with less regular
lamellar spacing parallel to the stretch direction than
that perpendicular to it, giving rise to the SAXS pattern
shown in Fig. 11(b).

$\vdash$— 0.5µ —$\dashv$

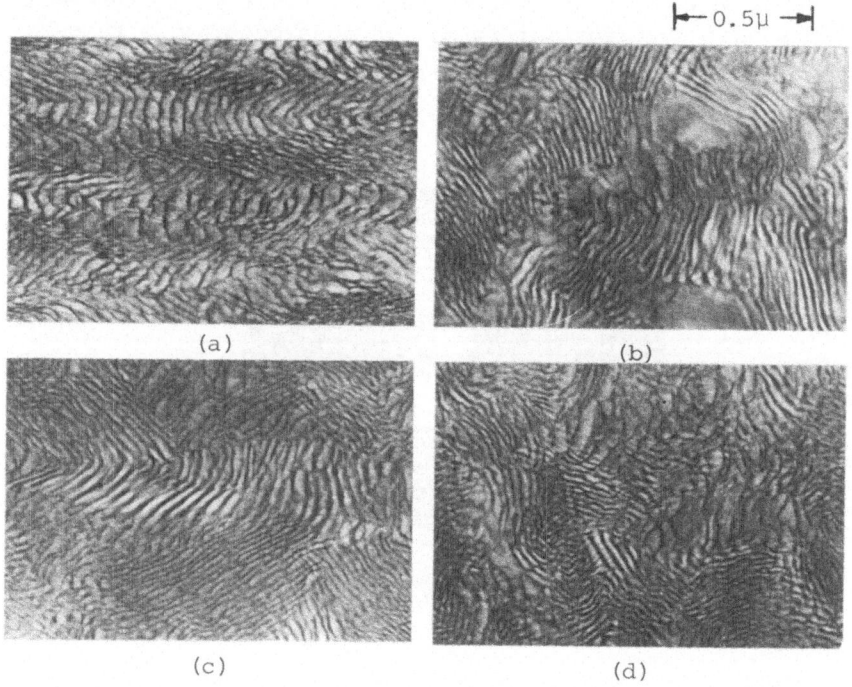

(a)   (b)

(c)   (d)

Fig. 12. Electron micrographs of SBS-PS specimens released
from 500% elongation and (a) left in unstretched state at
room temperature for a few days, and (b), (c) and (d) annealed at 60 , 88 and 116 $^{\circ}$C, respectively, all for 5 hr. Sectioning was made parallel to stretch direction and normal
to the film surfaces. The stretch direction is horizontal.

Figs. 13(a) and 13(b) show the changes of equatorial
SAXS intensity distribution around the first-order maximum
for various durations of healing at room temperature and at
an elevated temperature of 60 $^{\circ}$C, respectively. The SAXS
intensity distribution from unstretched (original) and
stretched (500% elongation) specimens are also included for

comparison. These figures demonstrate the recovery of the
diffuse SAXS intensity distribution of the stretched specimen to relatively sharp first-order maximum of the original
specimen. At room temperature no significant recovery can
be seen up to 4 hr, while at 60 °C almost complete recovery
except for a slight shift of the maximum to higher scattering angles, can be obtained in a duration as short as 10
minutes. When the duration is longer than 10 min., the
first-order maximum becomes more intensive and closer in
peak position to that of the original specimen, suggesting
not only the recovery of the domain structure in lamellar
spacing but also the increased regularity of lamellar
structure, i.e., the annealing effect.

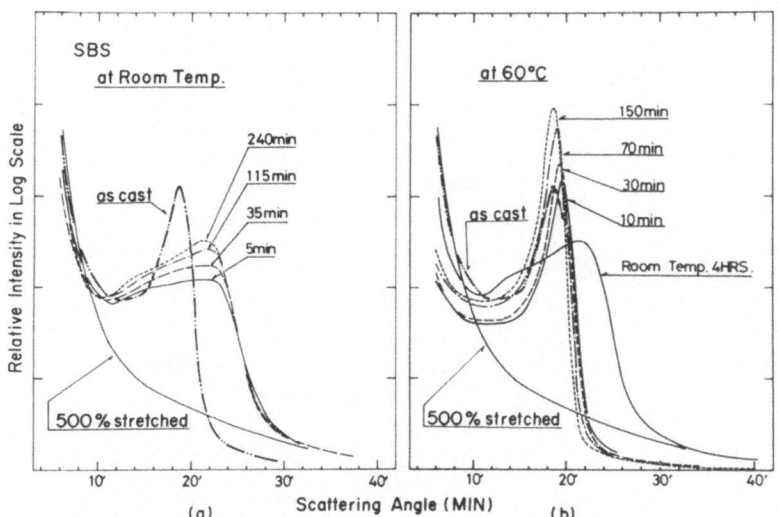

Fig. 13. Healing effect of SBS specimen released
from 500% elongation, investigated from the change
of SAXS intensity distribution with duration of
staying; (a) at room temperature and (b) at 60 °C.

Fig. 14 shows the temperature dispersion of isochronal
complex dynamic tensile modulus functions at a fixed frequency of 10 Hz for the SBS-PS specimen in unstretched and
stretched (330% elongation) states. The two temperature
dispersions around -100 and 90 °C in the unstretched state

may be assigned to the primary glass-transitions of the
polybutadiene and polystyrene domains. In the stretched
state, however, these loss peaks are broadened and shifted
to around -80 and 80 $^{\circ}$C, respectively. In addition, new
dispersion, as emphasized by a rapid decrease in $E'(\omega_o)$,
appears at around 40 $^{\circ}$C. The shift of the primary dispersion of polybutadiene matrix toward higher temperature may
be explained in terms of decrease of the free volume due to
internal stress arisen within the matrix. On the other hand,
the opposite shift of the primary dispersion of polystyrene
domain and the appearance of the additional dispersion may
be attributed to the fragmentation of the polystyrene lamellar domains.

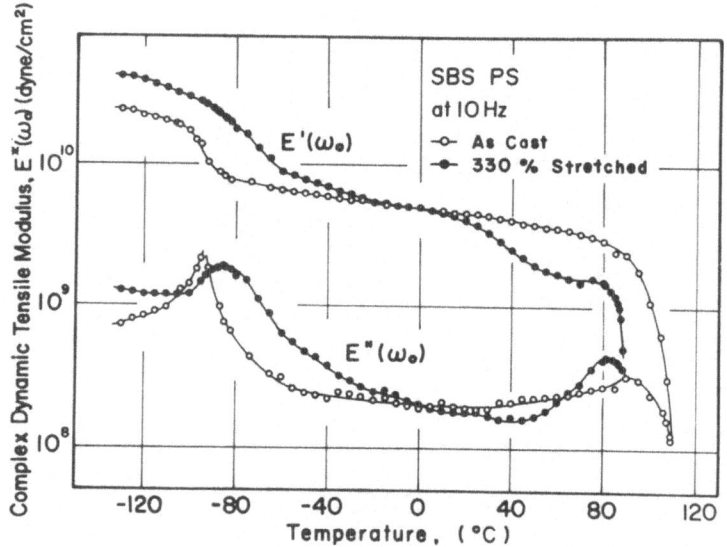

Fig. 14. Temperature dispersion of isochronal
complex dynamic tensile modulus function at a fixed
frequency of 10 Hz, observed for SBS-PS specimen at
unstretched and stretched (330% elongation) states.

The existence of an additional relaxation mechanism
has been noted by several authors for the heterogeneous
system of block and graft copolymers and been assigned to
a type of grain-boundary phenomena(20), i.e., interfacial

domain-boundary relaxation (3 -9), though the existence of
the additional relaxation mechanism has not always been
found as a discrete loss peak, but as a monotonous but
rather rapid decrease in E'(ω). Fragmentation of the polystyrene lamellae due to stretching must greatly increase
the specific surface areas of the styrene phase, and therefore the volume fraction of the interfacial domain-boundary
region. This results in the opposite shift of the primary
dispersion of polystyrene domains, as suggested by Braes
(21), as well as the appearance of the additional dispersion.

The fact that the structural reformation of the original lamellar domains from the fragmented ones can occur
even at a temperature as low as 60 $^{\circ}$C and a duration as
short as 10 min., as demonstrated in Fig. 13(b), may be
explained in terms of thermodynamic driving force, i.e.,
the fragmented system being associated with an excess free
energy relative to the original lamellar system owing to
(a) orientation of the polybutadiene chains (giving rise to
decrease entropy) and (b) enormous increase of the specific
surafce areas in the fragmented system (giving rise to increase interfacial energy). The structural reformation is
associated with the process of the enthalpy and entropy
relaxations. In the interfacial region, the volume fraction
of which is enormously increased in the fragmented system,
the polystyrene segments must be intermixed with polybutadiene segments, so that the polystyrene chains gain
enough mobility required for the structural reformation
even under conditions as mild as annealing at 60 $^{\circ}$C for 10
minutes.

As an application of the "plastic-to-rubber" and "its
inverse" transitions of the SBS type tri-block copolymers
to the development of a new composite material, lets one
discuss a new type of high impact polystyrene(HIPS) introduced by Aggarwal(22,23). Figs. 15 and 16 show electron
micrographs demonstrating the development of microcrazes
for two types of HIPS; one is ternary mixture of styrene-
isoprene di-block copolymer with corresponding homopolymers,
and the other is a commercial HIPS containing so-called
"salami sausage" type rubber domains dispersed in a styrene
matrix. For both materials, microcrazes are developed at
the boundaries of the rubber domains in the direction perpendicular to the impact strain. The development of micro-

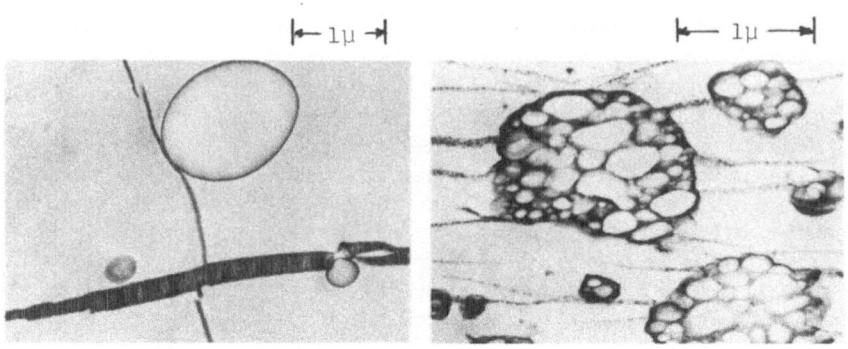

Fig. 15.   Fig. 16.

craze is a sort of micro-plastic deformation of the matrix
material accompanied by the formation of micro-fibrils. The
micro-fibril formation requires a great energy loss and is
believed as an origin of high impact strength of the material.

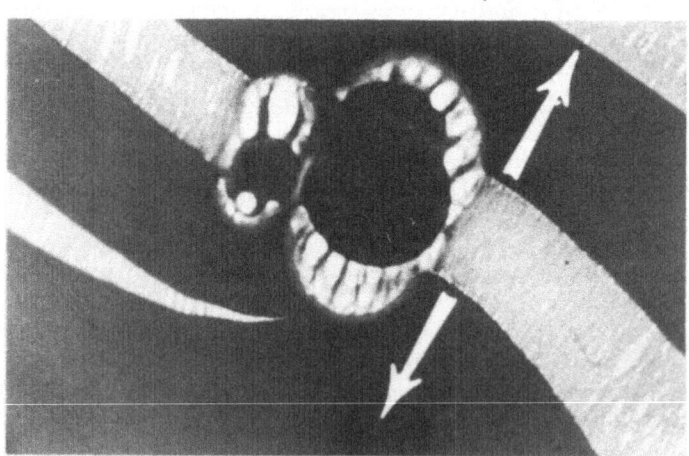

Fig. 17. Electron micrograph of a commercial
HIPS under a relatively high level of impact
stress. Arrows show the direction of the
impact tensile strain. (by courtesy of
Dr. S.L. Aggarwal, General Tire & Rubber Co.)

Fig. 17 shows electron micrograph demonstrating not
only the development of the microcrazes but also localized

deformation of dispersed domains of salami-sausage texture
for a commercial HIPS under a relatively high level of
impact stress. It is noted that the deformation of the
salami-sausage texture is concentrated in the rubber phase
but not in the styrene particles within the texture.

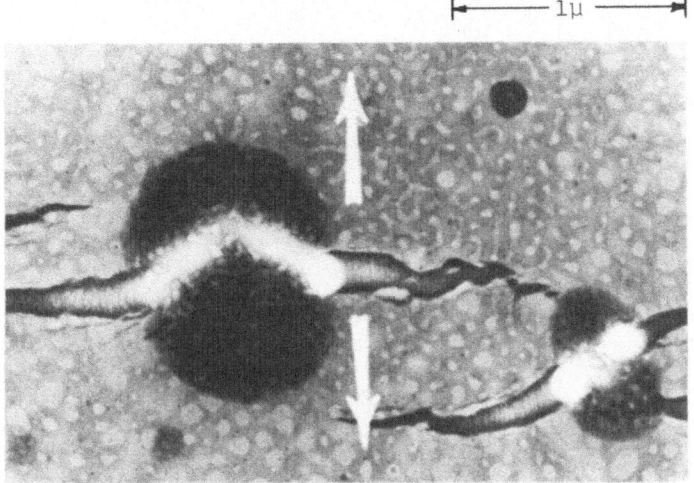

Fig. 18. Electron micrograph of a blend system
of 75/25 polystyrene/SBS tri-block under a
relatively high level of impact stress. Arrows
show the direction of the impact tensile strain.
(by courtesy of Dr. S.L. Aggarwal, General Tire
& Rubber Co.)

Fig. 18 shows electron micrograph of a blend system of
75/25 polystyrene/SBS tri-block copolymer, cast from toluene solution and subjected to a relatively high level of
impact tensile stress. The block copolymer is 40/60 S/B in
its fractional composition, and has an intrinsic viscosity
of 3.01(gm/dl)$^{-1}$. In contrast to the commercial HIPS in
Fig. 18, the most significance is that the microcrazes go
through the dispersed particles of the block copolymer. It
results in considerable local yielding and micronecking in
the block copolymer particle. The microdomain structure of
the particle is not of alternating lamellae of the two
components, but of styrene rods dispersed in a butadiene
matrix in the undeformed state. The domain structure of the
dispersed rods of styrene in the butadiene matrix still

Table I. Properties of Blends of SBS Triblock Copolymers with PS, HISP and Homopolystyrene (23)

Sample Description	% Diene Rubber in the System	Notched Impact Strength, ft lb/in.	Flexual Modulus x 10^{-5}, psi	Heat Distortion Temp. $^{\circ}$C
Polystyrene-1	0	0.2	4.9	92
Polystyrene-2	0	0.25	4.7	101
HIPS-1	3-5	1.0	3.46	75
HIPS-2	8-12	1.7	2.1	79
50/50 PS/SBS Triblock Blend. Block Polymer - 60/40 S/B Contents.	20.0	0.9	2.25	80
75/25 PS/SBS Triblock Blend. Block Polymer - 40/60 S/B Contents; [η] = 3.01 (gm/dl)$^{-1}$	15.0	7.5	3.00	87
75/25 PS/SBS Triblock Blend. Block Polymer - 40/60 S/B Contents; [η] = 1.62 (gm/dl)$^{-1}$	15.0	1.0	———	———

gives rise to the plastic-to-rubber transition, in some
extent, associated with fragmentation of the polystyrene
rods and with considerable loss of mechanical energy (19).
Besides the microcrazes of fibril formation, which are initiated at the boundaries of the rubber-rich block copolymer
particles and are developed, in a usual way, in the polystyrene matix, additional development of micronecking associated with the plastic-to-rubber transition within the
block copolymer particles must greatly contribute to improve
the impact strength of this type of new material (23).

ACKNOWLEDGMENTS

The authors are deeply indebted to Dr. S.L. Aggarwal,
vice president and director of the research division, The
General Tire & Rubber Co., Akron, Ohio, who furnished the
electron micrographs in Figs. 17 and 18 and discussed the
problem.

LITERATURES CITED

1) D.J. Meier, ACS Polymer Preprints, 15, 171 (1974); Polymer Colloquim Preprints, Kyoto, p.83, 1977.

2) E. Helfand, Accounts of Chemical Research, 8, 295 (1975);
   Macromolecules, 9, 879 (1976).

3) M. Shen and D. Kaelble, J. Polymer Sci., B8, 149 (1970).

4) T. Soen, T. Ono, K. Yamashita, and H. Kawai, Kolloid-Z.
   u.Z.Polymere, 250, 459 (1972).

5) S.L. Aggarwal, R.A. Livigni, L.F. Marker, T.J. Dudek, in
   "Block and Graft Copolymers", J.J. Burke and V. Weiss
   eds., Syracuse University Press, Syracuse, N.Y., 1973.

6) D.G. Fesko and N.W. Tschoegl, Intern. J. Polym. Mater.,
   3, 51 (1974).

7) T. Soen, M. Shimomura, T. Uchida, and H. Kawai, Colloid
   & Polymer Sci., 252, 933 (1974).

8) G. Kraus and K.W. Rollmann, J. Polymer Sci., Polym. Phys.
   Ed., 14, 1133 (1976).

9) G. Akovali, J. Diamant, and M. Shen, J. Macromol. Sci.-
   Phys., B13, 117 (1977).

10) T. Hashimoto, K. Nagatoshi, A. Todo, H. Hasegawa, and
    H. Kawai, Macromolecues, 7, 364 (1974).

11) T. Hashimoto, A. Todo, H. Itoi, and H. Kawai, Macromole-
    cules, 10, 377 (1977).

12) T. Hashimoto, A. Todo, H. Itoi, and H. Kawai, ACS Poly-
    mer Preprints, 18, 295 (1977).

13) A. Todo, T. Hashimoto, and H. Kawai, J. Appl. Cryst.,
    11, 40 (1978).

14) J.F. Henderson, K.F. Grundy, and E. Fischer, J. Polymer
    Sci., C16, 3121 (1968).

15) J.F. Beecher, L. Marker, R.S. Bradford, S.L. Aggarwal,
    J. Polymer Sci., C26, 117 (1969).

16) E. Fischer and J.F. Henderson, J. Polymer Sci., C26, 149,
    (1969).

17) G. Akovali, M. Niinomi, J. Diamant, and M. Shen, ACS
    Polymer Preprints, 17, 560 (1976).

18) M. Fujimura, T. Hashimoto, and H. Kawai, Rubber Chem. &
    Tech., 51, 215 (1978).

19) T. Hashimoto, M. Fujimura, K. Saijo, H. Kawai, J. Diamant,
    and M. Shen, Advances in Chemistry Series, ACS, in press.

20) N. Saito, K. Okano, S. Iwayanagi, and T. Hideshima, in
    "Solid State Physics", H. Ehrenreich, F. Seitz, and D.
    Turnbull, eds., vol.14, Academic Press, New York, 1963,
    p.458.

21) J. Bares, Macromolecules, 8, 244 (1975).

22) S.L. Aggarwal and R.A. Livigni, ACS Polymer Preprints,
    18, 212 (1977).

23) S.L. Aggarwal, paper presented at the U.S.-Japan Joint
    Seminor on Elastomer, Univ. of Akron, Akron, Ohio,
    October 17-21, 1977.

A NEW LOOK AT THE GLASS TRANSITION

J.J. Aklonis

Department of Chemistry
University of Southern California

A.J. KOVACS

Centre de Recherches sur les Macromolecules
CNRS-ULP, 67083 Strasbourg (France)

INTRODUCTION

Almost all theoretical papers on the glass transition in polymers or in low molecular weight glass formers, for that matter, fall into one of two broad categories. The first group (1) uses the methods of thermodynamics to relate the temperature or, more precisely, temperature range in which glass transition phenomena occur to other properties, either thermodynamic (such as changes in the heat capacity or coefficient of thermal expansion associated with the transition) or pseudo-molecular (such as hole formation energies and flex energies). To be sure, these attempts have had both successes and failures, the shortcomings often arising from the inability of such theories to treat the very complex time dependence associated with glass transition phenomena.

Work which falls into the second category (2-6) is directed at explaining or modeling the complicated time dependence. Here, too, various degrees of success have been realized but, since in these treatments the temperature or temperature range associated with glass transition phenomena is assumed to be known, little information concerning molecular factors and T_g is derived from such techniques. The work described below belongs to this latter category; one should realize, however, that a complete <u>understanding</u> of

glass transition phenomena will not arise until the concerns
of both of these schools are treated in a single coherent
framework (7).

FEATURES OF GLASS TRANSITION PHENOMENA

In this discussion we will be concerned with the rather
complex and varied types of behavior exhibited by materials
in the glass transition range. Several of the most interesting aspects of such behavior will be described briefly
below.

Effect of Experimental Rate

Figure 1 represents the behavior of glass formers subjected to constant cooling rate investigations (at this
point, all figures should be considered schematic; however,
they actually present calculated behavior based on the
model which will be described below). Such experiments
start at temperatures above the glass transition range and
then subject the sample to a constant rate temperature
decrease (for example, cooling rate = q = -1.0 units) while
monitoring an extensive thermodynamic variable of state,
such as the volume or enthalpy. (Although virtually all
considerations in this paper will be presented in terms of
isobaric volume changes, generalization to other appropriate
situations is straightforward). The response of the system
is volume contraction along the liquid equilibrium line at
high temperatures which is followed by positive departure
from this line at lower temperatures; at still lower temperatures, another sensibly straight line with small slope is
observed which corresponds to glassy behavior. The extrapolated intersection of these two straight lines is often used
to define the glass transition temperature T_g.

It is well known, however, that when a slower cooling
rate (q = -.01 units, for example) is employed in an otherwise identical experiment, the value of T_g observed varies
considerably from that determined with the faster cooling
rate as is shown in Figure 1. These experiments and many
others unequivocally show that glass transition phenomena
are time dependent. From this point of view then, the
concept of a <u>glass</u> transition temperature is very different
from an ordinary thermodynamic transition temperature since

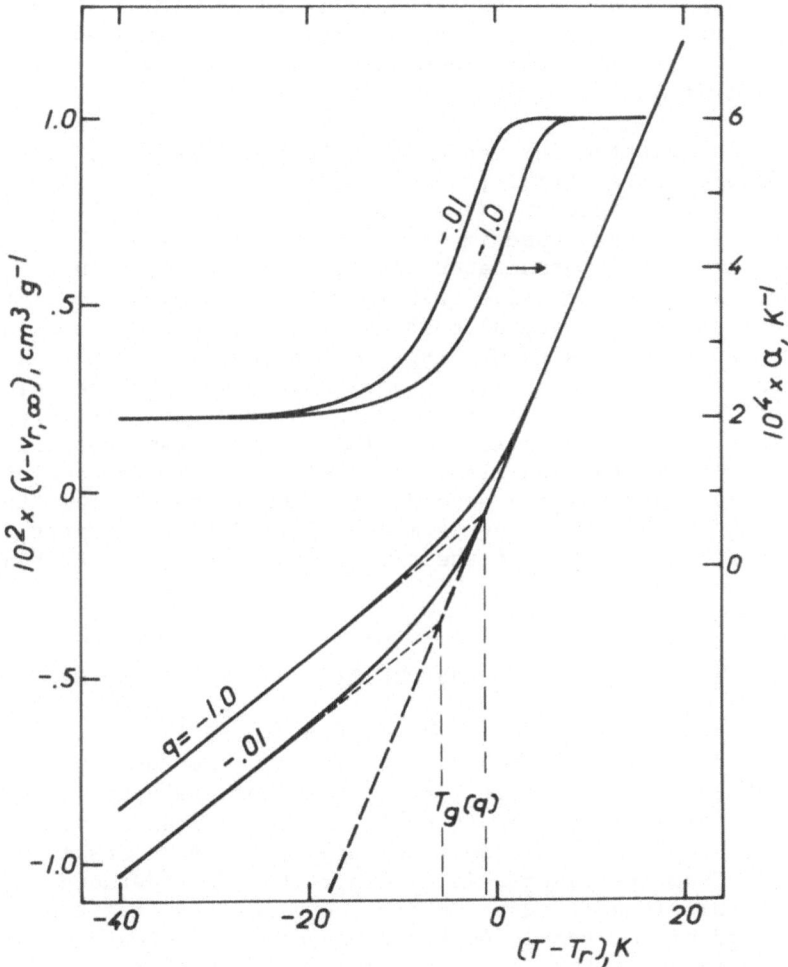

Figure 1.   Typical behavior of glass formers in constant
cooling rate experiments.   Cooling rate = q = dT/dt;
$v_{r,\infty}$ is the equilibrium volume at the reference temperature,
T_r, and α is the isobaric coefficient of thermal expansion.

T_g is <u>intrinsically</u> a function of the rate of the experiment and should properly be designated as $T_g(q)$ as in Figure
1.   It should be emphasized that the time dependence is an
inherent feature which is intimately involved in glass

transition phenomena and has nothing to do with carrying out
the experiment "too quickly" so that thermal gradients or
other irreversible experimental complications would influence
the results.

Also plotted on Figure 1 is the behavior of α, the
coefficient of thermal expansion for each of these experiments. It is seen that the value of α changes rather
abruptly for each experiment, from the high value
(6.0×10^{-4} K^{-1}) associated with the liquid state to a
considerably lower value (2.0×10^{-4} K^{-1}) indicative of the
glassy state. A true discontinuity in α, a second derivative of the Gibb's free energy, may be associated with a
second-order thermodynamic phase change; however, not all
discontinuities in α correspond to thermodynamic transformations as will be explained below. In constant cooling
rate experiments, the glass transition behavior resembles
that associated with a second-order phase change, and the
spacing of the two α curves measured at different rates is
further illustration of the time dependence of glass transition phenomena.

ASYMMETRY

One of the cooling isobars from Figure 1 is reproduced
in Figure 2 along with behavior observed upon heating glassy
samples at constant rate. First consider the upper curves;
these represent cooling ($q = -1.0$ units) to the temperature
$T = T_r - 10°$ and subsequent heating at the same absolute
rate. From the figure, it is clear that considerable
hysteresis is observed. In fact, one notices that the
behavior of the sample upon heating does not even approximately resemble the cooling response. In such a system,
where marked irreversible features are obvious, it is
difficult to see how purely thermodynamic arguments can be
correctly applied.

Rate dependence associated with glass transition
phenomena is also apparent here. For example, the lower
three curves depict realistic experiments starting at
equilibrium at $T = T_r - 10°$ with three different heating
rates. (No particular significance should be associated
with equilibrium as the starting point of these experiments).

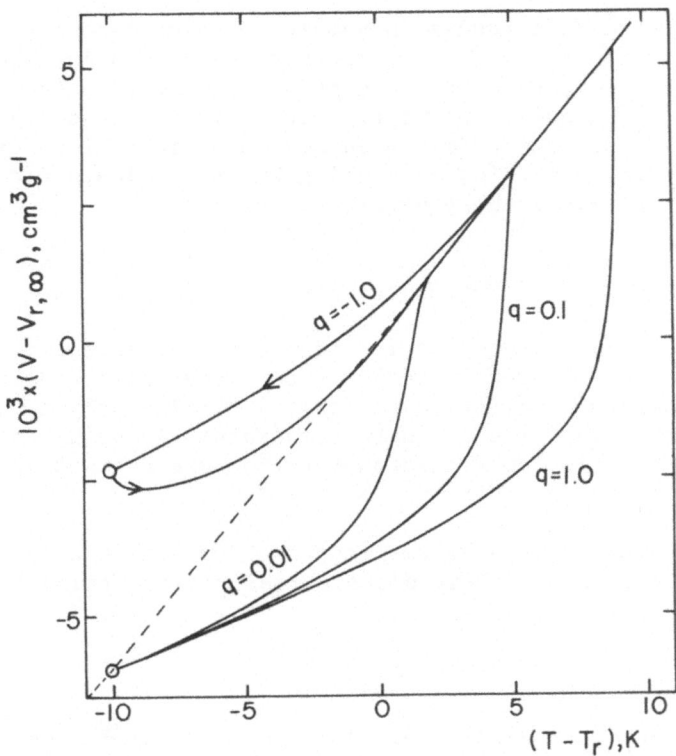

Figure 2. Typical behavior of glass formers in constant heating rate and constant cooling rate experiments.

The shapes of such heating isobars are more complicated than those of cooling isobars; after a delay period of varying length, these curves very rapidly approach the equilibrium liquid line. Thus, in marked contrast to cooling, heating behavior is reminiscent of a first-order thermodynamic phase change which, at the transition temperature, exhibits a mathematical discontinuity in the volume, a first derivative of the Gibb's free energy. This rapid approach toward equilibrium is responsible for the well known C_p peak which is observed upon heating annealed samples through the glass transition region; similar peaks are not observed upon cooling.

Profound differences in sample behavior associated with changes in the sign of the perturbation influencing the system are indicative of the distinct asymmetry which is evidenced by glasses in the transition region; and the asymmetry is so pronounced as to involve behavior which mimics a second-order transition in cooling but resembles a first-order transition in heating experiments.

Non-Linear Response

In the laboratory, it is often more convenient to use discrete step changes in temperature rather than constant rate temperature variations. This is particularly true for dilatometric experiments where temperature jumps are easily realized by removing a dilatometer from one thermostat and placing it into another.

In Figure 3, the volume response for six such experiments is presented. Here δ the normalized departure from

$$\delta \ = \ \frac{V-V_\infty}{V_\infty} \tag{1}$$

equilibrium, is plotted versus log time. V and V_∞ are temperature dependent values of the volume and equilibrium volume respectively. The upper three curves show contraction where the sample has been rapidly cooled and the delayed approach to equilibrium involves positive values of δ while the lower three experiments show retarded expansion ($\delta < 0$) after rapid temperature increases. Here, too, asymmetry in expansion and contraction is clear. However, another feature, the intrinsic non-linearity of the system response, is also noticeable. Consider, for example, the two contraction curves for $\Delta T = -2.5°$ and $\Delta T = -7.5°$. For a linear system the response to the larger temperature jump would be merely three times the response to the smaller jump; such behavior is not observed here. To emphasize the non-linearity of the system, the circles on each curve mark the point where 10% of the initial perturbation has relaxed. For linear response, the circles on the expansion isotherms and those on the

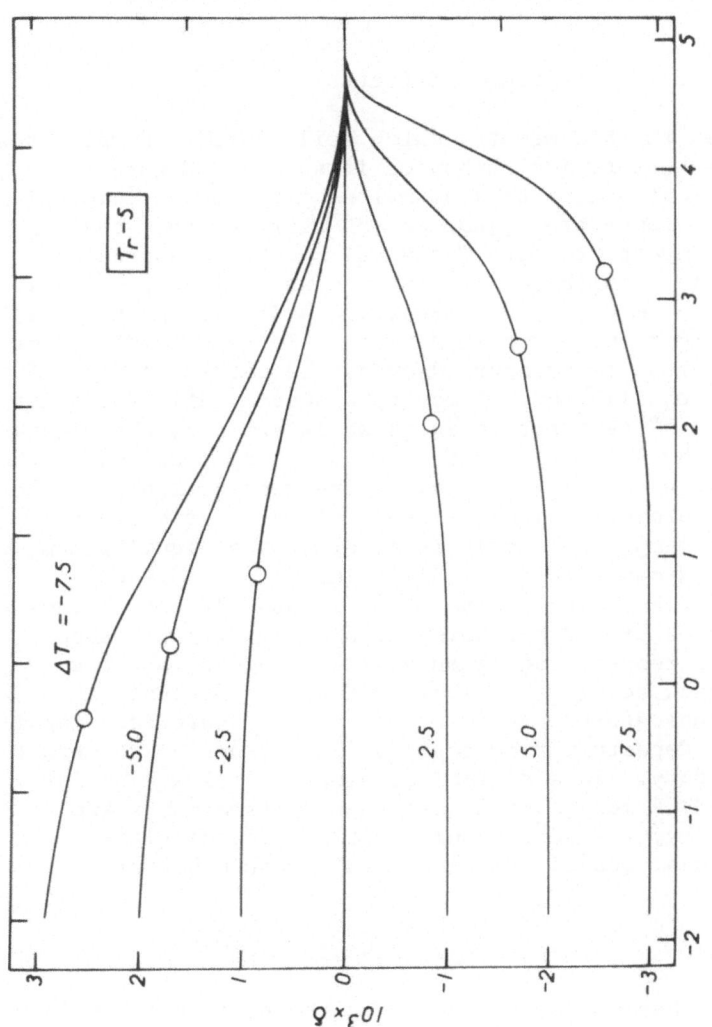

Figure 3. Simple approach behavior in the glass transition region. δ is plotted versus log t at the experimental temperature $T = T_r - 5°$. For contraction $\delta > 0$ and ΔT, the temperature jump, is negative; for expansion, $\delta < 0$ and $\Delta T > 0$.

contraction isotherms would fall on straight vertical lines.
For linear symmetric behavior, all six of the points should
fall on a single vertical line. The fact that all of these
points fall on a line which is far from vertical is clear
evidence of the pronounced non-linearity of response.

Memory Effects

Glasses exhibit memory, which still further complicates
their character; typical behavior is shown in Figure 4.
Here experiment A depicts a normal contraction experiment
induced by a temperature jump of -5° starting from the
reference temperature T_r. The usual monotonic approach
toward volume equilibrium is observed and $V(T = T_r - 5°)$ is
determined. The other four experiments (B through E) depict
multiple temperature jump situations. In experiment B, for
example, a -7.5° temperature decrease is used to perturb the
system from equilibrium. Recovery proceeds until the point
at which a 2.5° temperature increase is applied; the recovery
period was chosen so that this second T-jump resulted in the
sample being at $T = T_r - 5°$ and having its equilibrium volume
at this temperature. Thus, immediately after the second
temperature jump, the sample is apparently at equilibrium,
having been brought to its equilibrium volume ($\delta = 0$) at
$T = T_r - 5°$ via the two temperature jumps. If one follows
δ after such a procedure, however, one finds that δ spontaneously increases, taking on positive values, and then
decreases back toward equilibrium where $\delta = 0.0$, all at
constant temperature. As is clear in the figure, the magnitude of the departure from equilibrium depends on the exact
thermal history. In addition, much more complex behaviors
are observed if more complicated thermal treatments are
used. Clearly, any successful treatment of the glass
transition must quantitatively explain such behavior.

Complex τ_{eff} Behavior

From a thermodynamic viewpoint, the equilibrium volume
of a liquid is a unique function of temperature and pressure.
In Figure 1, the volumes for all such equilibrium states at
different temperatures combine to yield the equilibrium
liquid line, i.e., the straight line of high slope, a segment of which is dashed (extrapolated). Nevertheless, there
are other states of the system which are accessible and which

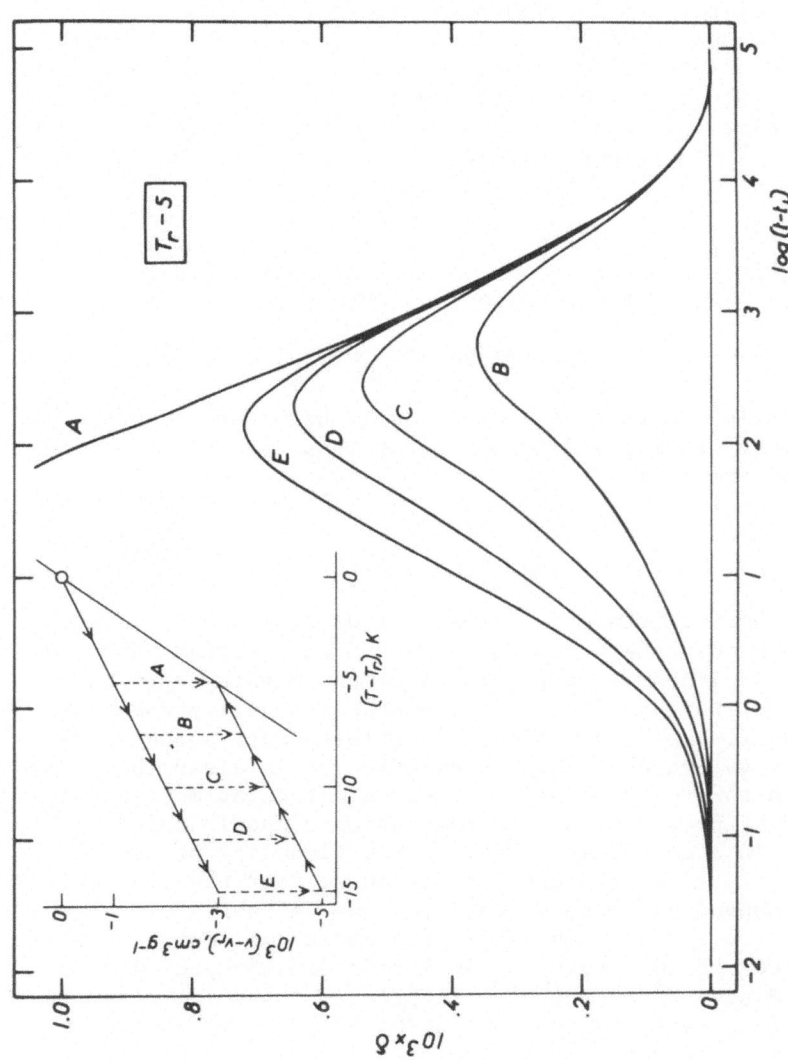

Figure 4. Memory behavior at $T = T_r - 5°$ evoked by multiple temperature jumps. Thermal histories are shown in the insert; t_l is the time at which the second T-jump is applied.

do not fall on this equilibrium line. Numerous such states
are represented in Figures 1 and 2, some having volumes
above the equilibrium liquid volume at the particular
temperature and others with volumes less than the equilibrium
liquid volume. To thermodynamically define such states in
the state space being considered, at least one additional
variable of state, often called an ordering parameter, must
be stipulated; this new variable measures how far from the
equilibrium liquid situation a particular state of the system is. Thus, in general, we may write

$$V = V(T,P,\zeta) \qquad (2)$$

where ζ is the ordering parameter. At equilibrium, ζ
becomes a unique function of T and P so that

$$V(T,P,\zeta(T,P)) = V(T,P) \qquad (3)$$

Such single ordering parameter models have been proposed in
the literature (2) and lead more or less directly to equations of the form

$$\frac{d\delta}{dt} = -\frac{\delta}{\tau(\delta)} \qquad (4)$$

for isothermal recovery. In such equations, $d\delta/dt$ is the
instantaneous rate of approach toward equilibrium which is
said to be linearly proportional to the magnitude of the
departure from equilibrium. The proportionality constant
τ, however, has interesting properties. Once again, considering volume changes, for example, it is clear that the
approach toward equilibrium involves molecular motion and
thus that the time scale of the response should depend,
perhaps in a complicated way, on the viscosity of the
system. It is well known, furthermore, that the viscosity
itself depends in a very sensitive way on the volume available to the system. Thus, loosely packed molecules will be
much more mobile than those in a more densely packed situation. Since volume recovery is precisely our interest,
these ideas lead to the conclusion that the viscosity or
intrinsic time scale of system must change during the
recovery experiment. To account for this, τ, the recovery
time, is defined as depending on the instantaneous state of
the system as measured by δ, i.e. τ is a function of δ.

Many forms have been suggested to express the actual quantitative dependence of τ on δ and these have been recently reviewed (4a).

Kovacs (2), however, chose to rearrange equation (4) considering $\tau(\delta)$ to be a dependent variable defined as τ_{eff}, the effective retardation time, which could be measured by analyzing simple approach response (response to a single T-jump starting with the system in equilibrium). Thus

$$\frac{1}{\tau_{eff}} = -\frac{1}{\delta}\frac{d\delta}{dt} \qquad (5)$$

which can be applied to both expansion and contraction data. A plot of log τ_{eff} versus δ for sixteen such experiments is shown in Figure 5. The right hand side is derived from contraction experiments, $\delta > 0$, while the left hand portion represents expansion behavior. To be sure, τ_{eff} behavior is a very sensitive measure of material response; in this figure the dashed line shows the δ dependence of τ as it is typically postulated in the literature (4a) for the single ordering parameter model. The great difference between the "theoretical" behavior for the single ordering parameter model and experimentally observed behavior is striking evidence of intrinsic shortcomings of the model.

Overall, we have indicated that materials exhibit complicated behavior in the glass transition region. Such effects as time dependence, asymmetry, non-linearity, memory and complex τ_{eff} behavior must be clearly explained by any successful theory.

RECENT THEORETICAL WORK

In the last several years, four reasonably similar treatments concerned with the kinetics of retarded recovery in glassy systems have been published (3-6). For purposes of this discussion, the differences between these treatments are not important; however, their essentially common fundamental features will be discussed within the framework of one of the models (4).

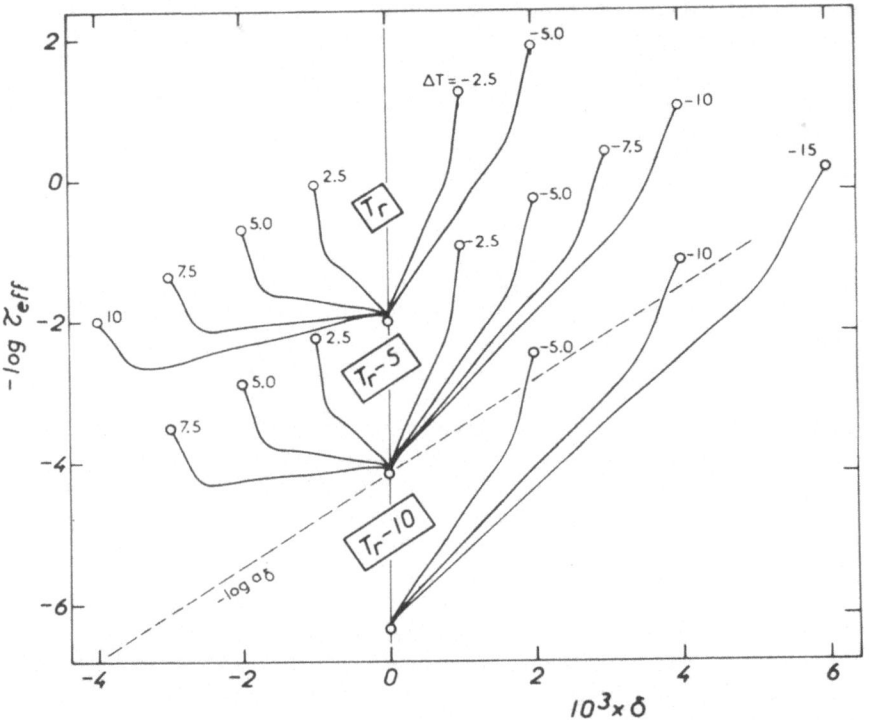

Figure 5. τ_{eff} (equation 5) behavior at three temperatures, T_r, $T_r - 5°$ and $T_r - 10°$ for simple approach expansion and contraction experiments in the glass transition region. ΔT, the value of the temperature jump used to perturb the system from equilibrium, is listed near each curve.

Single ordering parameter models suppose that any state of a glass can be uniquely defined by values of several thermodynamic parameters such as T and P, along with the value of the single ordering parameter. This assumption leads to an equation, like equation (4) for isothermal recovery, which implies that the entire approach to equilibrium is governed by one single equation in terms of one single ordering parameter.

However, a fundamental feature of polymeric systems which is obvious in virtually all experiments, is the breadth of the spectrum of response times. Thus $H(\tau)$, a distribution of viscoelastic relaxation times, usually spans many decades of time. This broad range results from the fact that localized molecular motions have much shorter relaxation times than the larger scale slower motions and that essentially all motions contribute to the relaxation process. In view of this situation, mathematical modeling of viscoelastic response always involves the distribution or partitioning of the response among relaxing elements with various fundamental response times. It is well known that a treatment of viscoelastic behavior in terms of a single Voigt element would meet with little success.

Nevertheless, this is precisely what is attempted when single ordering parameter models are applied to glassy systems. Here all of the response of the system is associated with a single fundamental recovery process, an approximation which, in our opinion, is too simplistic to be of very much use.

Following the viscoelastic analogy, then, one should partition the response among the various modes of recovery of the system. (The same conclusion is reached more rigorously in reference 4). In terms of ordering parameters, any particular state of a glass is then determined by T and P plus the values of N ordering parameters, one associated with each term in the partitioning. Equation 4 for isothermal recovery is modified to become

$$\frac{d\delta_i}{dt} = -\frac{\delta_i}{\tau_i(\delta)} \qquad (6)$$

where δ_i is the normalized departure from equilibrium associated with the i th mode of response. In equation (6), it should be noticed that $\tau_i(\delta)$ depends on the total departure from equilibrium δ and not on δ_i alone. Although δ_i dependence would be attractive here since the N equations would then decouple and simplify the mathematics, such functionality would stipulate that the retardation time of the i th mode depended only on features associated with that particular mode. These retardations times, however, are measures of molecular motion which actually occur in polymeric systems where intermolecular interactions are important. Thus, each motion is strongly influenced by the proximity of neighboring molecules and an appropriate measure of this influence is through δ and not merely δ_i. Therefore, we are forced, by the fundamental physical considerations of the model, to maintain the total δ functionality for each of the i retardation times; this dependence couples the N equations, which gives rise to many of the complicated features of glass transition behavior, as we will describe below.

At this point, then, we have roughly sketched a formalism by which recovery can be partitioned; we may now consider how partitioning influences τ_{eff} behavior. A very convenient method to represent partitioning is in terms of a distribution function like that shown in Figure 6. Here $g_i = \delta_{i,o} / \sum_N \delta_{i,o}$ is plotted as a function of $\log \tau_{i,r}$.

$\delta_{i,o}$ is the partitioned fractional departure from equilibrium associated with the recovery time $\tau_{i,r}$. (The second subscript on τ indicates the temperature, in this case, the reference temperature, T_r). This departure is induced by an instantaneous temperature jump starting with the system in equilibrium, $\delta_i = 0$, $1 \leq i \leq N$, and is measured at $t = 0^+$, i.e., before any recovery has occurred. This function represents the normalized intensity of the departure from equilibrium which should be associated with each retardation time $\tau_{i,r}$. On this figure two distributions are presented. The first represents the single ordering parameter model (the dot), and is shown merely as a bench mark. For this model there is no partitioning of response so the distribution function is a single point with intensity $g_i = 1.0$ at $\tau_{i,r} = 10.0$, which was arbitrarily chosen.

As we mentioned above, several expressions have been suggested to characterize the strucutral (δ) and thermal dependences of recovery times. In a recent review (4a), it was shown that, in a narrow temperature interval in the glass transition range, all of these functions reduce to that first

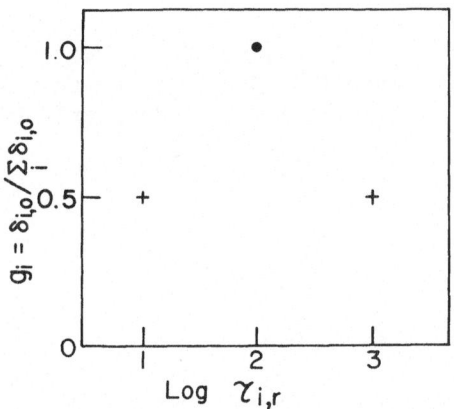

Figure 6. Distribution of normalized instantaneous departures from equilibrium, g_i, as a function of recovery times
τ_i. Two distributions are shown, the single ordering parameter model as a dot and a simple equal intensity partitioning as the two crosses.

suggested by Toole (8):

$$\tau_i(\delta,T) = \tau_{i,r} \exp(-\theta(T-T_r)) \exp-(1-x)\frac{\theta\delta}{\Delta\alpha} \quad (7)$$

Here $\tau_{i,r}$ is the value of the i th recovery time at the
reference temperature at equilibrium, $\theta = E_a/RT$ where E_a
is the activation energy for the process, $\Delta\alpha = \alpha_\ell - \alpha_g$
where α_g and α_ℓ are coefficients of thermal expansion of
the glass and liquid respectively and x is a factor which
mediates the influence of structure and temperature on the
recovery times. $x = 1$ indicates purely thermal dependence

while x = 0 gives purely structural dependence. Since we
are more interested, at this point, in treating glass transition phenomena qualitatively rather than quantitatively, we
will adopt the reasonable values of all of the parameters
listed below. To be sure, variations in these parameters
from material to material should be expected.

TABLE I
Values of Parameters Used in the Calculations
Characteristic Values for Polymeric Glasses

$$\alpha_g = 2.0 \times 10^{-4} \text{ K}^{-1}$$

$$\Delta\alpha = 4.0 \times 10^{-4} \text{ K}^{-1}$$

$$\theta = 1.0 \text{ K}^{-1}$$

$$x = 0.4$$

$$v_{r,\infty} = 1.0000 \text{ cm}^3 \text{ g}^{-1}$$

With this information, one can determine the τ_{eff}
behavior for the single ordering parameter model; this is
shown in Figure 7 as the dotted line. Again, as was noted
above, the straight continuous line for both expansion and
contraction is far from what is observed experimentally
(Figure 5).

In Figure 6 an additional distribution function is also
represented. Here, the departure from equilibrium is
partitioned equally ($g(\tau_i)$ = .50) among two recovery modes,
τ_1 = 10 and τ_2 = 1000. In Figure 7, the τ_{eff} behavior
associated with one expansion and one contraction experiment
is also shown. This behavior is calculated by subjecting
the model to an instantaneous temperature jump and then
using equation 6 to calculate the time dependence of δ
through recovery; subsequently equation 5 is applied to the
results to obtain the τ_{eff} behavior shown. Also shown here
(the partially dashed lines) is the single ordering parameter behavior for these two retardation times.

A short discussion will make the behavior resulting
from the partitioned distribution function (Figure 6) clear.

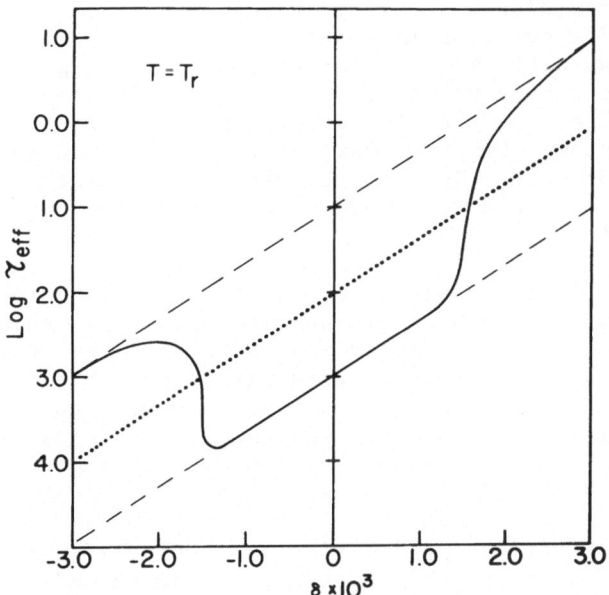

Figure 7. τ_{eff} behavior associated with the distribution functions shown in Figure 6. The dotted line corresponds to the single ordering parameter model and the solid line represents the equal intensity partitioning case.

In the contraction experiment, a rapid temperature decrease causes departure from equilibrium; half of this departure is associated with one retardation time and half with the other. Clearly, the initial approach toward equilibrium will be dominated by the faster process, i.e., that with the

shorter retardation time. In fact, this is exactly what
happens since the calculated τ_{eff} behavior approximately
follows the single ordering parameter model behavior for the
shorter retardation time at large values of δ. However, in
the region of 50% recovery, all of the departure from
equilibrium controlled by this more rapid process has
recovered; and in order to recover further, the system must
shift to the slower recovery process, as is shown in the
figure. The behavior in expansion is entirely analogous,
starting with the more rapid process dominating τ_{eff} at
large departure from equilibrium and terminating via the
slower process in the final approach toward equilibrium,
after a rather abrupt transition from the rapid to the slow
recovery process. (One of the recent treatments (6) mentioned above would give more complicated behavior in such
an analysis).

    Thus, it is possible to partition the departure from
equilibrium to avoid single straight line τ_{eff} behavior.
One may then ask if complex τ_{eff} behavior like that observed
experimentally can be generated with a suitable distribution
function. The answer here is yes; moreover, the particular
distribution function which was used to generate all of the
"schematic" data presented in Figures 1 through 5 is shown
in Figure 8.

    This distribution function is particularly simple; it
involves two boxes, the shorter time one extending over two
decades of log τ and containing only about 15% of the intensity (area) of the function and the longer time box also
spanning two decades but being much more intense (85%).
From our initial investigation of volume recovery of several
polymeric systems, it is our impression that small changes
in the shape or extent of the distribution function will be
necessary to treat particular materials but that the gross
features of this function seem to be generally valid.

            FEATURES OF GLASS TRANSITION PHENOMENA AND THE MODEL

    The five most prominent features of glass transition
phenomena were discussed above and will now be rationalized
in terms of the model sketched in the preceding section.

    The complex τ_{eff} behavior has already been discussed.
In fact, it was just this feature of the glass transition

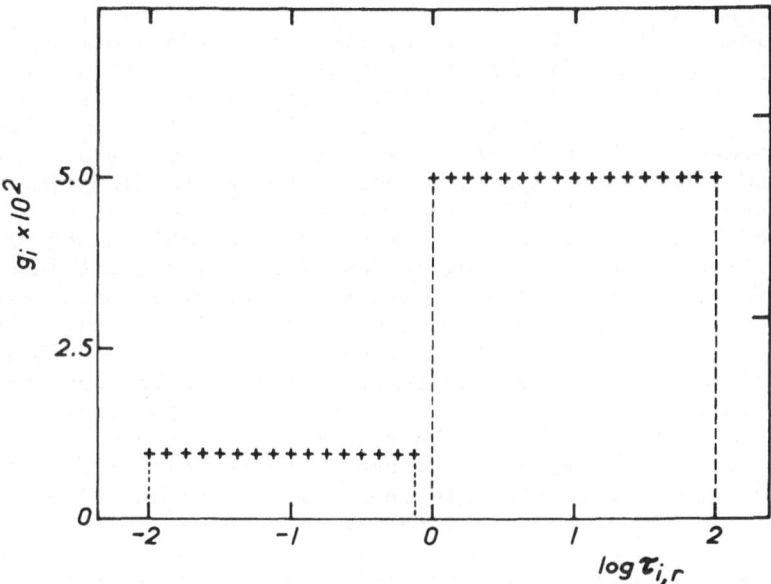

Figure 8. Realistic distribution of normalized instantaneous departures from equilibrium g_i as a function of recovery times τ_i. This distribution corresponds to the data shown in Figures 1 through 5.

which was used to determine the appropriate distribution function given in Figure 8. It should be noted here that the complexity results from the interplay between the structural dependence of the individual retardation times, as expressed in equation 7, for example, and the partitioning of departure from equilibrium as represented by the distribution function.

The time dependence associated with glass transition phenomena is also easy to understand. As is obvious, all of the equations presented to describe the phenomena are functions of time; overall time dependence is thus not surprising. From a more intuitive viewpoint, however, this model is in accord with the usual interpretation of behavior like

that shown in Figure 1. For the sake of discussion, let us
consider the constant rate temperature decrease cooling
experiment to be a series of minute T-jump-induced contraction experiments. For the more rapid cooling rate,
starting at higher temperature, the small T-jump displaces
the system from equilibrium slightly, but under these conditions the contraction period is long enough so that
equilibrium is reestablished before the imposition of the
next temperature jump. As this sequence of events progresses, the temperature decreases, the retardation times
increase and, eventually, the time between temperature
changes is not sufficient to allow the reestablishment of
equilibrium before the application of a subsequent T-jump.
At this point, the system behavior trajectory departs from
the equilibrium liquid line, causing the "glass transition".
With a slower cooling rate, the periods for the reestablishment of equilibrium between temperature jumps are longer, and
the equilibrium line is followed to lower temperatures
before glassy response is observed. In slower experiments,
then, the model predicts that the glass transition occurs
at lower temperatures due to the interplay of the time scale
of the experiment and the kinetics of recovery.

We will next discuss the intrinsic asymmetry associated
with this transition. First, however, a particular view of
the general relationship between theoretically derived distribution functions and experimentally observed response
functions will be presented, with the distribution of viscoelastic relaxation times $H(\tau)$ and the stress relaxation
modulus $E(t)$ serving as a specific example. Mathematically,
these functions are related by the expression:

$$E(t) = \int_{-\infty}^{\infty} H(\tau)\, e^{-t/\tau}\, d \ln \tau \qquad (8)$$

When a sample is instantaneously strained, the total stress
is partitioned among the various modes of relaxation
according to the function $H(\tau)$. At any time, t, the relaxation modulus $E(t)$ is computed by multiplying $H(\tau)$ by the
time dependent exponential $e^{-t/\tau}$ (often referred to as the
kernel) and then integrating. At short times, the kernel
(which varies from 1.0 to 0.0) has a value of 1.0 over the
entire range of $H(\tau)$, so that the short time value of the
modulus is no more than the total area of the distribution
function. As time increases, i.e., as relaxation proceeds,

the time kernel moves toward larger values of τ, thus trun-
cating the short relaxation time portions of the distribution. Now integration gives less area and the calculated
value of $E(t)$ decreases monotonically with increasing time.

The situation is similar in the present context of
$\delta(t)$ (the response) and g_i (the distribution function),
except for one very important difference. Consider the
portion of Figure 9 marked "contraction". The distribution
function is shown by solid lines. The experiment consists
of quickly cooling this sample from a high temperature to
the experimental test temperature, say $T_r - 5$, as shown in
the insert; at the start of the experiment, then, the sample
is in state A. The distribution function, however, as represented by the solid lines is depicted at equilibrium, i.e.,
at state B. Since the time scale of response is structure
(δ) dependent, the distribution function at the start of the
experiment must actually be located at shorter retardation
times and this is shown by the dashed line spectrum in the
figure. Once again, system response is followed by the
progression of the time kernel, which is again an exponential, shown as the dotted curve which moves from left to
right across the figure. However, as the kernel and the
displaced distribution begin to overlap, indicating recovery,
the volume of the system departs from state A and begins its
trajectory toward state B, i.e., equilibrium. But in state
B, the distribution is represented by the solid line; so
during the recovery, as the kernel moves from left to right,
the distribution function also moves in the same direction.
This motion is represented in Figure 9 by the arrows, one
associated with the time kernel and the other with the displaced distribution function. Thus the protacted response
for contraction as shown in Figure 3 results from the kernel
chasing the receding distribution function. (The essential
difference between this situation and stress relaxation is
that here the distribution function actually moves on the
log τ scale during recovery, while in stress relaxation
$H(\tau)$ is usually considered to be fixed).

A complementary situation exists in expansion. Here
the initial state, immediately after the T-jump, is represented by state C in the insert of Figure 9, and the initial
volume is below the equilibrium volume. In this case, then,
the initial position of the distribution function in Figure
9 will be at longer retardation times (dashed lines) than is

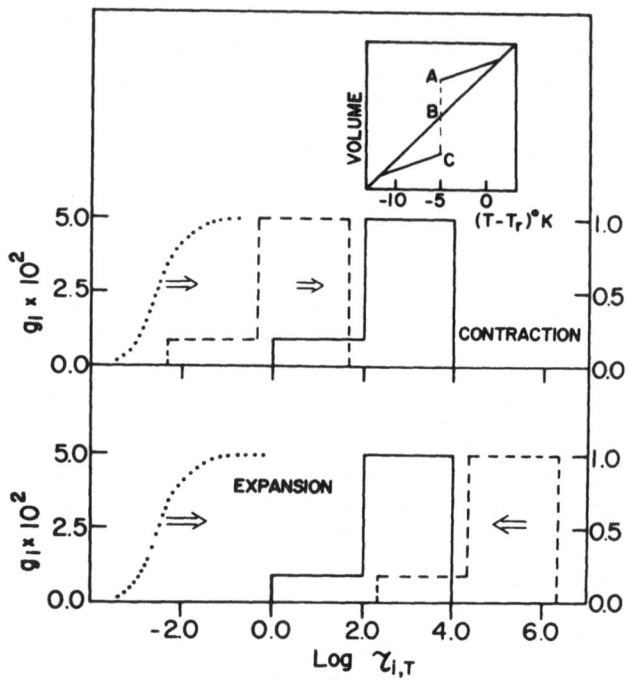

Figure 9.   Fundamental explanation of asymmetry in glass
transition phenomena as interpreted by the model.

the case in equilibrium (solid lines).   Here it should be
clear that a very long plateau at constant δ should precede
any recovery since the time kernel must travel for a long
distance until it begins to interact with the displaced
distribution.   This plateau is obvious in Figure 3.   As
recovery commences, the volume increases from point C toward
B, which again causes the displaced distribution to move

toward the equilibrium position indicated by the solid lines. Now, however, the distribution moves from right to left, sweeping under the advancing time kernel and causing a very rapid approach toward equilibrium once it starts.   In heating experiments, it is just this rapid approach toward equilibrium which gives rise to the C_p and α peaks which are sometimes given thermodynamic significance.   It should be clear, however, that here such peaks are not associated with thermodynamic transitions but instead arise straight-forwardly from the kinetics of recovery which is structure dependent.   Thus, the marked asymmetrical aspects of glass transition behavior are satisfyingly explained in terms of the structural dependence of the distribution of recovery times.

The non-linear aspects of recovery result   from the same feature of the model.   However, it is informative to consider this feature from a different viewpoint.   Equations (6) and (7) can be combined to yield

$$\frac{d\delta_i}{dt} = -\frac{\delta_i}{\tau_{i,T} \, a_\delta} \qquad 1 \leq i \leq N \qquad (9)$$

where a_δ contains all of the structural dependence from equation 7.   Although this equation is very simple, it cannot be integrated directly because of the a_δ term in the denominator.   One can, nevertheless, define a reduced time u

$$u = \int_0^t \frac{dt'}{a_\delta} \qquad (10)$$

which transforms equation (9) into a directly integrable form, so that

$$\delta_i(u) = \delta_{i,o} \, e^{-u/\tau_{i,T}} \qquad (11)$$

The system response given by this set of equations is linear in terms of the reduced time u, since multiplication of the perturbation by a constant will merely multiply $\delta_{i,o}$ by the same constant, causing the response $\sum_N \delta_i(u)$ to be multiplied by the identical factor.   The reduced time, u, is the fundamental time scale of the system; and, in terms

of this intrinsic time scale, a glassy system does, in fact,
behave linearly. The experimentally observed non-linearity
is associated with the relationship between the intrinsic
time scale u and the laboratory time scale t as defined in
equation (10). This relationship is graphically presented
in Figure 10 where log u is plotted versus log t for several
of the simple approach experiments of Figure 3. From this
plot one can see that u and t differ significantly, $u \geq t$
for contraction experiments and $u \leq t$ for exapnsion.

The final feature of glass transition behavior to be
discussed is memory; and this results from the incorporation of a distribution into the model. In Figure 11,
normalized values of δ_i versus $\tau_{i,r}$ (recovery times at
equilibrium) are plotted for various states of recovery for
a contraction experiment. In this particular representation,
the function does not move along the log τ scale as recovery
progresses since the values of τ represented are those at
equilibrium; on the other hand, the function does change
shape as recovery progresses, since $\delta_i(t)$, which is time
dependent, is plotted rather than $g(\tau_i)$. As expected, the
departure from equilibrium associated with shorter recovery
times decreases first. This recovery is followed by similar
decreases associated with the longer time scale processes
and, after sufficient time, equilibrium is reestablished
($\delta_i = 0$, $1 \leq i \leq N$). Expansion experiments would also be
depicted in the same way, but here all of the values of
$\delta_i(t)$ would be negative.

As was explained above, memory behavior may be observed
when a partially equilibrated contracting sample is rapidly
heated to a temperature where it adopts its equilibrium
volume, i.e., when $\delta = 0.0$. In terms of partially equilibrated states like those in Figure 11, this merely amounts
to adding a negative δ component via a temperature increase
equal to the positive δ component at that particular state
of recovery. However, although the overall departure from
equilibrium is zero, this state has been realized by two
temperature jumps which result in a partitioning from
equilibrium like that depicted in Figure 12 for the pairs
of temperature jumps which were shown in the insert in
Figure 4. For each of these states, the apparent equilibrium situation results from compensating negative and positive departures from equilibrium. However, in Figure 12 it
is clear that the $\delta_i < 0$ components are all associated with
rapid recovery processes while those with $\delta_i > 0$ have slower

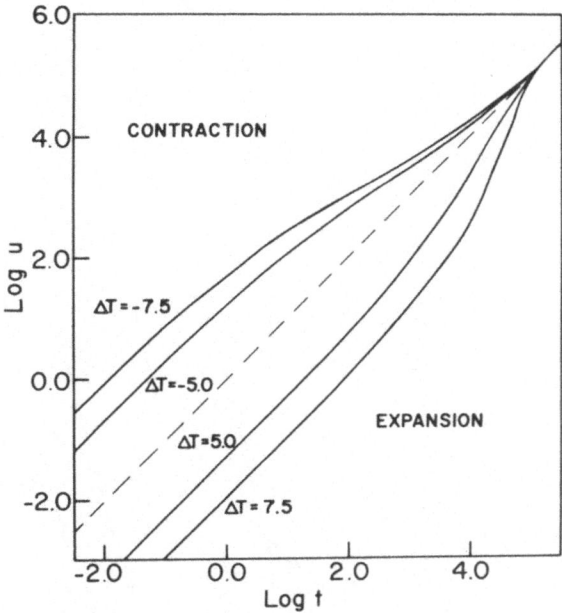

Figure 10. Relationship between the experimental time t and the intrinsic time u (equation 10) for several of the simple approach experiments depicted in Figure 3.

recovery. Thus, the negative components will recover first, initially causing δ to increase (volume spontaneously departs from its equilibrium volume at constant temperature). This increase will be followed by a maximum and then a decrease in δ back to the fundamental equilibrium state which is characterized not by $\Sigma\ \delta_i = 0$ but rather by $\delta_i = 0$,

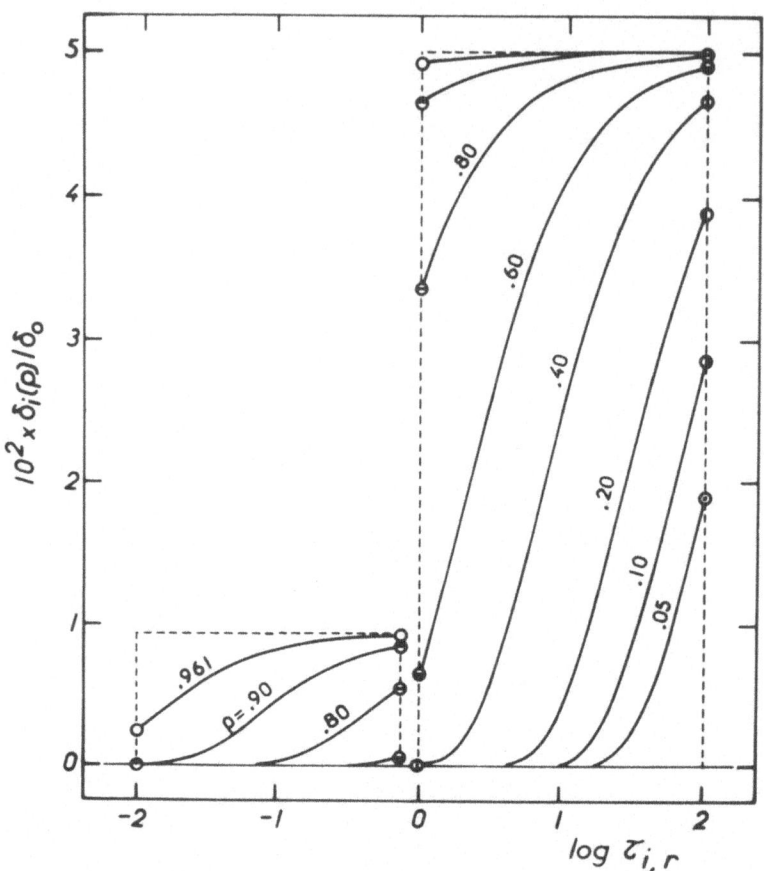

Figure 11.   Normalized departures from equilibrium versus
retardation time during recovery for simple approach experiments. $\rho = \delta(t)/\delta_0$ with specific values of ρ listed near
each curve.

$1 \leq i \leq N$.   According to this analysis, the observation of
any memory in glassy systems is <u>very</u> strong evidence for
the existence of a distribution of mechanisms, all of which
contribute significantly to recovery.

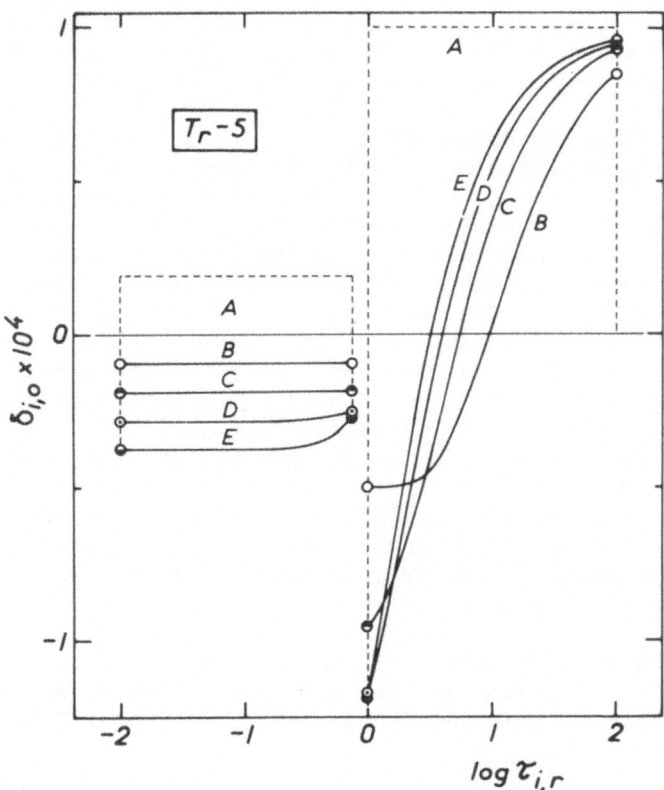

Figure 12.   Departures from equilibrium versus retardation
time for the multiple T-jump experiments detailed in the
insert of Figure 4; here the time is immediately after the
application of the second T-jump.

CONCLUSIONS

Although there are many features of glass transition
phenomena which appear very complicated and even perhaps
contradictory, virtually all of them are satisfyingly
explained in terms of a simple model which partitions

departures from equilibrium among a set of structure-depen-
dent recovery times. As mentioned previously, rigorous
treatments of the model from different viewpoints have
been presented elsewhere (3-6). In this paper we have
focused on a description of the behavior of the model for
glassy systems in general and we have not attempted to
analyze the behavior of any one particular glass.

Our focus has been the elucidation of specific features
of the model which give rise to the most important aspects
of glass transition behavior. Thus time dependence arises
naturally from a consideration of the molecular aspects of
the overall phenomena involved. The pronounced non-
linearity and asymmetry of behavior result from the structure dependency of the individual retardation times introduced in the model. In fact, our analysis shows that, on
a time scale appropriately compensated to take account of
structural dependence, non-linearity and asymmetry vanish.
The existence of a multiplicity of recovery times in the
model leads to memory, which is observed in real systems.
And finally, there are other features of glassy behavior,
such as the complex nature of τ_{eff}, which arise because of
a subtle interplay between the distribution of recovery
processes and the structural dependence of the recovery
times for these processes.

In view of the simplicity of this model and its
success in explaining so many features of glassy behavior,
we feel that further development of this area will be
significant.

REFERENCES

1. For example: T.G. Fox and P.J. Flory, J. Appl. Phys.,
    21, 581 (1950); R. Simha and R.F. Boyer, J. Chem. Phys.,
    37, 1003 (1962); J.H. Gibbs, J. Chem. Phys., 25, 185
    (1956); and J.H. Gibbs and E.A. DiMarzio, J. Chem.
    Phys., 28, 373 (1958).
2. A.J. Kovacs, Fortsch. Hochpolym. Forsch. (Adv. in
    Polymer Sci.), 3, 394 (1963).
3. O.S. Narayanaswamy, J. Amer. Ceram. Soc., 54, 491 (1971).
4. (a) A.J. Kovacs, J.M. Hutchinson, and J.J. Aklonis,
         The Structure of Non-Crystalline Materials, Ed.,
         P.H. Gaskell, Taylor & Francis, London (1977),
         p. 153.

   (b)   J.J. Aklonis, J.M. Hutchinson, A.R. Ramos, and
         A.J. Kovacs, J. Polymer Sci., in press.
5.   M.A. DeBolt, A.J. Eastel, P.B. Macedo, and C.T.
     Moynihan, J. Amer. Ceram. Soc., 59, 16 (1976).
6.   Ryong-Joon Roe, J. Appl. Phys., 48, 4085 (1977).
7.   R.E. Robertson, J. Polymer Sci., C, F.P. Price
     Memorial Symposium, in press.
8.   A.M. Toole, J. Amer. Ceram. Soc., 29, 240 (1946).

NEW METHODS OF PRODUCTION OF HIGHLY ORIENTED POLYMERS BY SOLID STATE EXTRUSION

Michael P.C. Watts, Anagnostis E. Zachariades
and Roger S. Porter
Polymer Science and Engineering Department
Materials Research Laboratory
University of Massachusetts
Amherst, Massachusetts 01003

INTRODUCTION

The development of thermoplastics of high tensile modulus has been the subject of intense research in recent years. This can be achieved by introducing a high degree of chain orientation and extension by solid state deformation of semicrystalline polymers by the processes of drawing (1) and extrusion (2).

Work in this laboratory originated in studies of the crystallization of an oriented high density polyethylene (HDPE) melt using an Instron capillary rheometer (3). It was found that these polyethylene rods had a remarkably high modulus (up to 70 GPa) (4). The process was modified to a solid state extrusion by Capiati, et al. (5), in that melt-crystallized billets of HDPE were extruded below their melting point through conical dies of draw ratio up to 40. Mead (6), Kojima (7) and Perkins (8) studied the effect of varying processing conditions on properties and observed the range of materials that may be prepared by this method.

We wish to review here a series of developments made in this laboratory that have extended the range of the solid state extrusion technique for HDPE and also allows the extrusion of materials previously found to be difficult to process by conventional solid state extrusion. The properties of materials produced by these unconventional processing methods will also be described. The topics to be discussed are:

Section 1. Extrusion of High Density Polyethylene

(a) Conventional Solid State Extrusion
(b) Split Billet Extrusion
(c) Push-Pull Extrusion

Section 2. Processing of "Unprocessable" Polymers

(a) Conventional Solid State Extrusion
(b) Powder Extrusion
(c) Solid State Coextrusion

Section 3. Properties of Oriented Ultrahigh Molecular Weight
Polyethylene Prepared using the Powder and
Coextrusion Techniques

Section 4. The Use of Ammonia as a Plasticizer for
Nylon 6 in Solid State Extrusion

Summary

Section 1. Extrusion of High Density Polyethylene

(a) Conventional Solid State Extrusion

The processing and properties of HDPE rods produced by solid
state extrusion will be briefly described; for more detail, the
reader is referred to the original papers (6-8) and an extensive
review by Zachariades, et al. (9). The polyethylene used was a
duPont grade Alathon 7050 Mw = 57,000, Mw/Mn = 3. Melt crystallized billets were prepared by melting at 160°C and subsequent
crystallization under the combined effects of temperature and
pressure. The billets were extruded through a conical brass die,
with an included entrance angle of 20°, in a 1 in. bore Instron
capillary rheometer.

Figure 1 shows that as the draw ratio increases the apparent
viscosity increases and hence the extrusion rate drops rapidly.
The rapid increase in viscosity is thought to be related to the
strain hardening observed in a conventional tensile test at high
elongation (10). The extrusion rate is promoted by high extrusion
pressure, however, there is an upper limit to the pressure.
When the pressure is increased much above 0.23 GPa, stick-slip extrusion occurs. At an extrusion pressure of 0.23 GPa, the extrusion stops when the viscosity increases sufficiently that the extrusion force cannot overcome the resistance to extrusion. During slip in stick-slip extrusion, a spiral fractured extrudate is
produced that is similar to spiral products found in shear fracture of polymeric melts during extrusion (12). The flow profile
in the solid state extruded rods is a deep shear parabola (10, 13,
24) (Figure 2) which suggests that stick-slip may arise from
shear fracture of the extrudate.

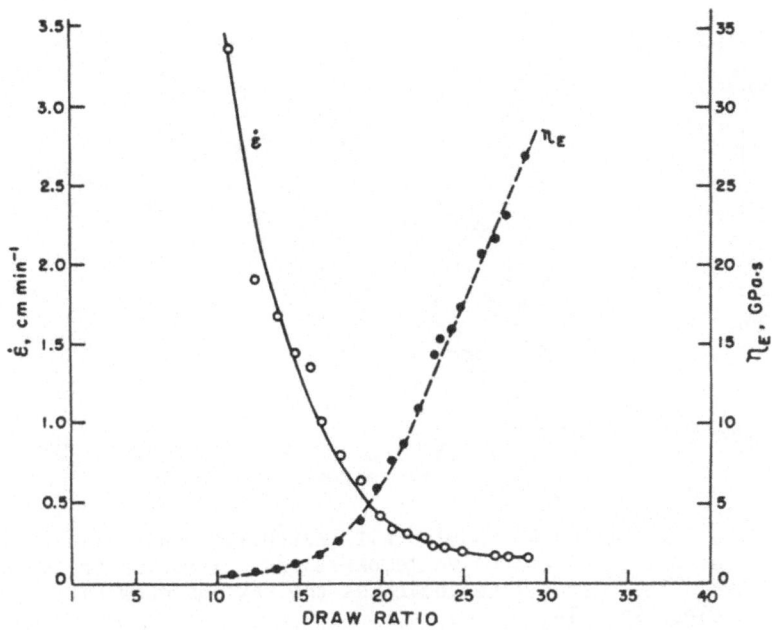

Figure 1: The relations between viscosity, extrusion rate and draw ratio for low molecular weight HDPE (Alathon 7050).

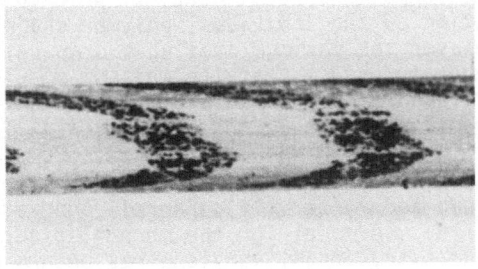

Figure 2: Shear flow profile for low molecular weight HDPE (Alathon 7050) at extrusion draw ratio 24.

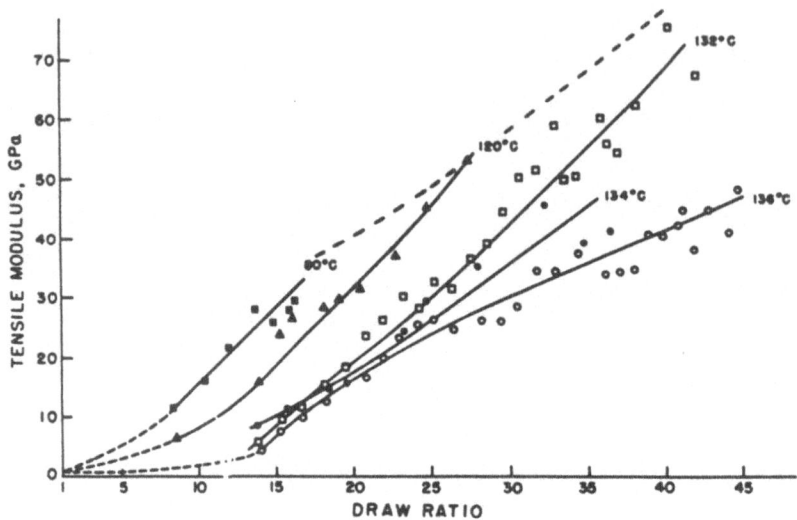

Figure 3: Variation of modulus with extrusion draw ratio at a series of different extrusion temperatures, extrusion pressure 0.23 GPa. The dashed line indicates the maximum draw ratio attainable. Ref. 14.

The range of moduli and draw ratios available by this process is shown in Figure 3. The moduli are measured at 0.1% strain and a strain rate of 2.0×10^{-5} sec^{-1}. The lower the extrusion temperature, the higher the modulus, presumably because of greater efficiency of deformation. Efficiency of deformation is the ratio of the deformation of the individual polymer chains to the bulk deformation. However the maximum available draw ration is also dependent on temperature such that the maximum modulus is observed for samples extruded at 132°C and a draw ratio of 40. The physical properties of the extruded rods are given in Table 1, and a comparison with other high modulus materials is made in Table 2. Detailed analyses of the properties of highly oriented HDPE have been made by Mead and Porter (14), Farrell and Keller (15), Ward and coworkers (16). A model for the morphological changes in highly drawn and extruded polyethylene has been proposed by Peterlin (17), alternate models have been proposed by Clark and Scott (18) and Porter (19). The key to the high modulus of these morphologies seems to be in the high modulus of the PE unit cell (~ 240 GPa) (20,48). The PE chain crystallizes in the trans-trans conformation so that only bending and extension of carbon-carbon bonds can occur, and the crystallized chain has a very high

TABLE 1

Comparison of Physical Properties of Ultraoriented HDPE
Rods with Single Crystal Material

Physical Property	Polyethylene Single Crystal	Ultraoriented PE Rod
Tensile modulus,GPa	285, 240[47,48]	70
Strain to fracture,%	6 (49)	3%
Tensile strength,GPA	13	0.4
Linear Expansion Coefficient, 10^{-5} $^{\circ}$C^{-1}	(a)-axis + 22[50] (b)-axis + 3.8 (c)-axis -1.2	Fiber axis -0.9 $\pm$ 0.1
Melting point, $^{\circ}$C	145.5[51]	139°C (10°C min^{-1} heating rate)
Crystallinities		85%
Crystal orientation function	1.0	0.996 $\pm$ 0.002
Birefringence,total	0.059[52]	0.0637 $\pm$ 0.5015

TABLE 2

Presently Achievable Polymer Tensile Properties

Material	Tensile Modulus (GPa)		Tensile Strength (GPa)
	Small Crystals	Bulk Samples	
Polypropylene	42 (48)	22 (53)	0.93 (53)
Polyethylene	240 (48)	70 (54,55,56); 110* (60)	0.60 (57)
Polystyrene		4.5 (33)	0.08 (33)
Carbon Steel SAE 1020		208 (58)	0.62 (58)
Aluminum		69 (58)	0.17 (58)
Glass Fiber		69-183 (59)	0.69 (58)

*PE fibers grown from sheared solution.

modulus in the chain direction. Clearly the solid state extrusion technique is very efficient in producing high modulus HDPE,
and hence the interest in extending the range of the extrusion
technique.

(b) Split Billet Extrusion

The first departure from conventional extrusion came with
the invention of the split billet technique (21). A melt crystallized billet of HDPE (Alathon 7050) was split in half longitud-
inally (Figure 4); the two halves are then placed together and extruded through a conventional conical brass die. The flow pro-
files in Figure 2 were obtained by marking the inside surface of
the split billet. Figure 5 shows the effect of splitting the
billet on the extrusion rate. Initially the conical part of the
billet is extruded so the draw ratio of the extrudate steadily
increases until the cylindrical part of the billet is reached and
"steady state" extrusion occurs with the extrudate having a constant draw ratio. Clearly the split billet extrusion proceeds at
a higher rate and, perhaps more importantly, the maximum draw
ratio obtainable without fracture was increased. This is thought
to be due to stress relief from the addition of a free surface in
the center of the billet (21). The products from the split billet
extrusion are indistinguishable from conventionally extruded Alathon 7050 rods processed under the same conditions. The increased
draw ratio available using this technique is illustrated in Figure 6, a plot of extrusion draw ratio against extrusion temperature.

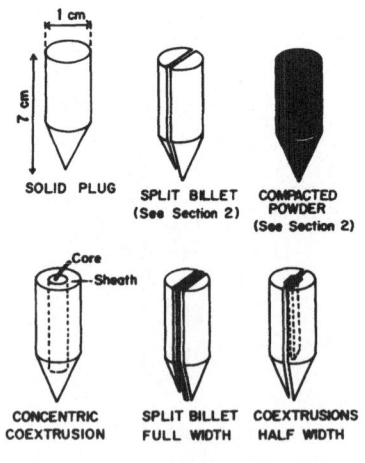

Figure 4: Schematic of the different sample geometries used in
this study.

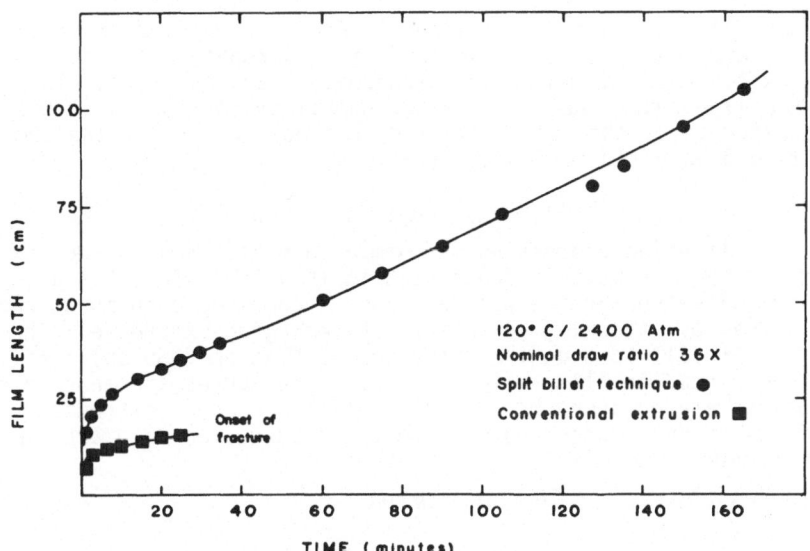

Figure 5: A comparison of the extrusion rate for a split billet and solid plug extrusion. Ref. 21.

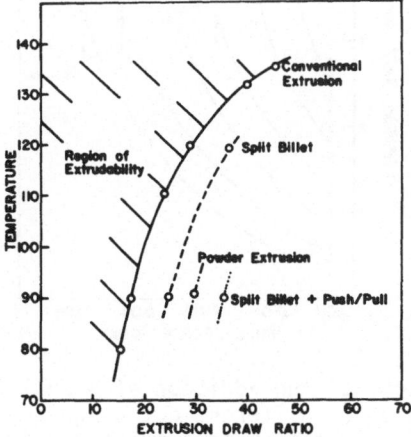

Figure 6: Processing map for Alathon 7050, extrusion pressure 0.23 GPa. Lines connect points of maximum draw ratio available at given temperature.

At a given temperature, there is a maximum draw ratio that can be successfully extruded. The lines in Figure 6 join up these points of maximum draw ratio. Above 138°C the extrudate melts at the die exit, hence, a region of extrudability can be drawn. The dashed line shows the limit of extrudability for the split billet technique, i.e. the split billet allows new regions of the temperature/draw ratio map to be studied.

(c) Push-Pull Extrusion

Pulltrusion processing is common in metals and polymer composites. The extrusion process used in this laboratory has a possibility of extruding the polymer under a combined push force and pull load by attaching weights to the emerging extrudate. The effect of pull load on the extrusion of Alathon 7050 split billets has been studied (11). Figure 7 shows the marked increase in extrusion rate produced by adding a pull load to the extrudate. The effects of push force and pull force on extrusion rate are found to be superimposable.

Push force = 3.0X (Pull Force)

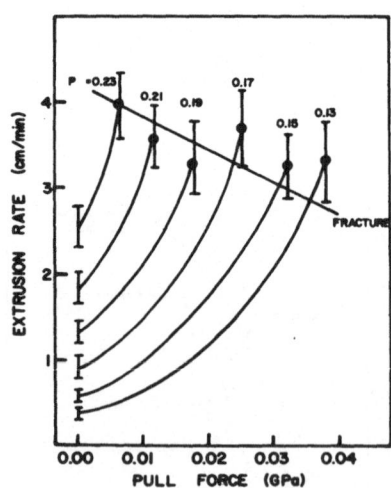

Figure 7: The effect of the addition of a pull force on extrusion rate at a series of different extrusion pressures P. Addition of the pull load increases the extrusion rate until fracture occurs in the die; marked ● on the graph. The polymer is a split billet of Alathon 7050, extrusion draw ratio 36, temperature 120°C.

The factor 3 arises from the hydrostatic transmission of forces
through the polymer in the die. When too large a pull force is
used, the extrudate fractures at the base of the conical part of
the die (Figure 7). From the relationship between push and pull
forces, the pressure at the base of the conical part of the die may
be calculated (0.08 GPa). This is similar to the tensile strength
of the extrudate (0.04 GPa) at the extrusion temperature. Although
the extrusion proceeds under compression, the material is being
elongated in the die, hence the correspondence between the pressure
at fracture and tensile strength.

In conventional extrusion the maximum pressure that can be
applied without stick-slip is around 0.23 GPa and so this provides
a "process dependent" restriction on extrusion rate and the maximum draw ratio available. In push-pull extrusion, forces up to
the tensile fracture of the material can be applied so the true
maximum draw ratio for the material at a given temperature can be
obtained. The limit of temperature and draw ratio detected in
these studies has been added to the extrudability map (Figure 6)
showing the maximum regions of the temperature/draw ratio map
that may be studied. Clearly the push-pull process greatly
extends the range of the extrusion technique. A series of samples prepared at the same draw ratio and temperature, but with
different combinations of push and pull, have the same modulus,
i.e. mechanical properties are independent of the applied pressure.

Section 2.   Processing of "Unprocessable" Polymers

This section deals with semicrystalline thermoplastics that
cannot be processed by the techniques described earlier. Perkins
and Porter (22) have reviewed the solid state deformation of
polymers in detail and describe the numerous reports of solid
state extrusion. Aharoni (23) has reported that a number of
polymers may be solid state extruded to high draw ratio (> 10) by
the conventional process. These include HDPE, poly(ethylene oxide), poly(4-methyl pentene-1). Polypropylene is also readily
extrudable (24). However, there are many other polymers that
would be attractive if they could be obtained in high draw, particularly the established fiber-forming polymers such as the nylons and poly(ethylene terephthalate). The maximum extrusion
draw ratio that has been reported for nylon 6 is 5 (25). This
has been attributed to the onset of strain hardening at much lower
extensions than polyethylene (10). Ultrahigh molecular weight
polyethylene is also of interest as a way of improving the mechanical properties.

We will describe some new developments which allow these materials to be extruded to higher draw ratios. Details of the polymers described in this section are given below.

TABLE 3

Polymers Used in This Study

Polymer	Producer	Grade	$\overline{M}_w$
High density Polyethylene	DuPont	Alathon 7050	59,000
Ultrahigh molecular weight Polyethylene	Hercules	UHMW 1900	5×10^6
Nylon 11	Rilsan	Besno	28×10^3
Poly(ethylene terephthalate)	Goodyear	VFR 5041 VFR 5610	IV = 1.04 IV = 0.6
Polystyrene	Monsanto	HH 101	333×10^3
Poly(vinylidene fluoride)	Kureha, Japan	—	—

(a)   Conventional Solid State Extrusion

We have attempted to extrude nylon 11, poly(ethylene tereph-thalate) (PET) and ultrahigh molecular weight polyethylene (UHMWPE).   The billets were prepared by slow cooling from the melt.   The results are summarized below.   Nylon 11 was extruded up to draw ratio 5 in the presence of silicone oil lubricant.   Both high molecular weight PET and UHMWPE could not be extruded at this draw ratio.   Use of the split billet and push-pull techniques did not assist the extrusion.   For the purposes of this discussion these polymers will be described as "unprocessable".

TABLE 4

Conditions for Conventional Solid State Extrusion of
Some Fiber Forming Polymers at Draw Ratio 5

Polymer	T_m, °C	T_{ext}, °C	P_{ext}, GPa	Ext. Rate cm min^{-1}
UHMWPE	134	120	0.23	S-S
Nylon 11 no lubricant	194	180	0.23	S-S
Nylon 11 silicon oil lubricant		120	0.16	2.0
PET IV = 1.04	270	260	0.4	S-S
IV = 0.6	270	250	0.23	1.0

S-S indicates that stick-slip occurred

(b) Powder Extrusion

First we discuss the extrusion of a compacted powder. Powder processing is used extensively for metals and ceramics. There are only a few investigations and applications to polymers. PTFE (26) is processed as a granular powder in molding and extrusion processes. The compacted powder is heated to the gel state, shaped, and allowed to solidify in the mold. A patent claiming the use of powdered UHMWPE for processing above the melting point (27) and a paper describing the compaction of these powders have also appeared (28).

This development involves the extrusion of a compacted powder billet prepared <u>and</u> extruded <u>below</u> the melting point. The powder is loaded into the 3/8" diameter barrel of an Instron capillary rheometer that is plugged at one end. A moderate pressure (~ 0.23 GPa) is applied to compact the powder into a coherent billet. This billet is then extruded in the conventional manner giving a translucent coherent strand.

A sample of powdered Alathon 7050 was prepared by precipitating a solution in xylene into methanol. The compacted billet was extruded at draw ratio 30 and at $90^\circ C$, conditions that are unattainable in conventional extrusion (6) (Table 5). This datum can be added to the extrudability map (Figure 6) and shows that using a powder is not as efficient at promoting the extrusion as the push-pull process. There is also some loss of properties in the powder extrudate (Section 3).

UHMWPE is available commercially as a powder and if it is prepared as a compacted billet as above, it may be extruded up to DR 24 at $128^\circ C$ whereas a solid plug or split billet cannot even be extruded to DR 5, a major change in extrudability (30). The properties of these products are described in Section 3.

Conventional extrusion is thought to proceed by extension of an entangled molecular network. There is no molecular network between powder particles so the particles are thought to be deformed by interparticle adhesion and friction. Moreover, it is this lack of molecular entanglement that improves the extrudability.

As well as offering a way of handling difficult materials, this technique provides access to entirely new morphologies (29). A large quantity of solution grown single crystal mats of Alathon 7050 were prepared from xylene solution. These were packed into the rheometer and extruded. The presence of the highly structured single crystal gave a further improvement in the extrusion rate. Details of the processing and properties are the subject of a separate publication (29).

TABLE 5

Conditions for Extrusion of UHMWPE by the Powder and Coextrusion Processes

Sample	Preparation			Extrusion		
	Press.GPa	Temp.°C	Draw Ratio	Press.GPa	Temp.°C	Rate $\frac{cm}{min^{-1}}$
Alathon 7050 Powder Extrusion	0.23	90	30	0.23	90	0.06
UHMWPE Powder Extrusion	0.23	128	24	0.23	128	0.2
UHMWPE* Coextrusion Low Pressure Cryst.	0.15	180	5	0.15	120	2
UMWPE* Coextrusion High Pressure Cryst.	0.46	210	5	0.20	120	2

*Nylon 11 sheath was prepared by melting at 230°C and crystallized by cooling slowly under 0.1 GPa pressure.

(c) Solid State Coextrusion

The coextrusion processes are the second group of methods for processing "unprocessable" polymers. They are divided into split billet coextrusion and concentric billet coextrusion. In principle they involve the extrusion of the unprocessable component in a sandwich with a processable polymer, to promote extrusion. A form of substrate drawing is involved.

Split Billet Coextrusion. During the development of the split billet technique, it was realized that by placing a polymer film between the billet halves, the film would be extruded with the outside halves and ultradrawn thin film could be produced (Figure 4). 0.1 mm thick films of Alathon 7050 draw ratio 24 were produced at 120°C, a draw ratio that was unobtainable using a slit die. The film in the center of the billet is forced to follow the two outside halves because of the compression in the conical die and friction between the components.

This technique was then applied with poly(vinylidene fluoride) films to a maximum draw ratio of 7 at 120°C. PVF_2 was of interest because of its piezoelectric properties, the films had a moderate modulus, and complex melting behavior (31,32).

Alternatively a film of amorphous polystyrene was placed in a HDPE split billet and when extruded at 126°C to draw ratio 12 gave a thin oriented film. This film had a fibrous structure and is the first report of an amorphous polymer having a microfibril morphology (33).

These examples show how versatile the split billet coextrusion process is in producing oriented thin films.

Further insight into the coextrusion process has developed from work on amorphous polyethylene terephthalate films. Films of amorphous PET were prepared by melt pressing at $280°C$ and quenching in ice water. These were then extruded at $100°C$ in an Alathon 7050 split billet. Unoriented amorphous PET does not crystallize at a significant rate at this temperature (34). On extrusion through a draw ratio 7 die, films 0.5 mm and 1.0 mm thick emerged with draw ratio 7. However, a film 2.0 mm thick emerged with a draw ratio of 5, insufficient compressive forces were transmitted by the two polyethylene halves to the thick film. For the film and substrate to undergo the same strain, the stress in the film and substrate must be similar and the stress in film may be controlled by the cross sectional area.

Attempts to extrude 1.0 mm thick films of amorphous PET at draw ratio > 7 failed because of fracture at the edges of the film. All the film extrusions discussed so far involve films that are the same size as the flat surface of the split billet. A 1.0 mm thick film of PET was cut so that it was 1/2 the width of the billet surface and positioned in a groove in the center of the billet (Figure 4). Using this configuration, a coherent film of draw ratio 12 was obtained, nearly double the previous value.

This illustrates the second important feature of coextrusions. The flow profile of a full width Alathon 7050 film is a deep shear profile, the same as the split billet (21) (Figure 2), i.e. the edges of the film undergo severe shear stress. In contrast, a film that covers only the central part of the billet will undergo essentially extensional deformation. The shear strength is lower than the tensile strength, particularly in highly drawn materials, so the full size PET film undergoes fracture at the edges whereas the half size films do not and may be taken to much higher draw.

These examples illustrate the use of the split billet coextrusion process for film extrusion and show some of the factors that are important in optimizing the process.

Concentric Billet Coextrusion. The concentric billet extrusion uses a billet of processable polymer with a hole drilled in the center. A rod of "unprocessable" polymer is placed in this hole and the composite extruded (Figure 4). The concentric composite billet was first used to prepare "one polymer composites" in which the two components consist of two samples of HDPE of different molecular weight (35). Many features of these composites have been studied (36,37).

In the case of a concentric billet consisting of two different polymers, the outer sheath can act merely as a processing aid and/or polymer die. After extrusion it is removed to leave the core component.

The concept is illustrated by the processing of ultrahigh molecular weight polyethylene. The outer sheath was a melt crystallized billet of nylon 11 that could be extruded to draw ratio 5 at 120°C and 0.16 GPa using a silicone oil lubricant. Two samples of UHMWPE were prepared. The UHMWPE was melt crystallized in the rheometer at two different temperatures and pressures. The low temperature, low pressure conditions were chosen so as to prepare a lamella morphology; the high temperature, high pressure conditions to give a chain-extended product (38) (Table 5). The billets were machined so that they fitted inside the nylon 11 sheath. The composite was extruded at draw ratio 5 and 120°C. The nylon casing was removed to free the core which was a coherent fracture free rod. The presence of the outer sheath allowed UHMWPE to be extruded to a higher draw than is possible by conventional extrusion.

The nylon casing is extruded with the core; therefore there is no shear at the surface of the core. Also, the material at the center of the billet does not undergo shear deformation as much as the surface in contact with the rheometer walls, so again shear stresses are minimized and shear fracture (stick/slip) prevented. These features are the same as for split billet coextrusion. In addition, the sheath has significant tensile strength perpendicular to the chain direction, hence the core is held under a significant compressive force after the sample has left the die. This insures plastic rather than elastic deformation of the core. The importance of the tensile strength of the sheath is illustrated by an attempt to extrude an UHMWPE core and a low molecular weight HDPE sheath - the sheath fractured.

The possibility of using other casings so as to increase the draw ratio is under investigation, as is the possibility of using a molten core in a solid casing.

Section 3. Properties of Oriented Ultrahigh Molecular Weight Polyethylene Prepared Using the Powder and Coextrusion Techniques

The mechanical properties of oriented UHMWPE are used to indicate that polymers that require unconventional processing may also have unconventional properties. Details are given elsewhere (30).

Experimental

The mechanical properties of extruded rods were measured in tension in an Instron TTM tester at a strain rate of 2×10^{-5} sec^{-1}. The tangent modulus at 0.1% strain is recorded. The degree of crystallinity was measured on a Perkin-Elmer DSC 1B at a heating rate of 10°C min^{-1} using Indium calibration and a heat of fusion of 69 cal g^{-1} for polyethylene crystals.

Powder Extruded UHMWPE

The principal properties of an UHMWPE powder extrudate with draw ratio 24 are given in Table 6. The most obvious feature of the rods is that they are opaque whereas rods of Alathon 7050 from melt crystallized billets are transparent. The density of the powder extruded rod was 0.96 gcm^{-3}, a value consistent with the degree of crystallinity of 70%. This indicates that the opacity is a result of minute density fluctuations rather than the presence of gross voids. The tensile modulus is lower (15 GPa) than the modulus of Alathon 7050 melt crystallized, and drawn to a draw ratio of 24 (25 GPa). However, this is a substantial increase over unoriented material. Clearly significant orientation and chain extension has occurred during extrusion. Tensile strength was lower than Alathon 7050 but elongation at break was the same.

TABLE 6

Mechanical Properties of UHMWPE Products

Sample	Draw Ratio	Tensile Modulus/ GPa	Tensile Strength/ GPa	Elonga- tion at Break %	Degree of cryst. %
Alathon 7050 Solid billet	24	25	0.25	3	85
UHMWPE Powder extrusion	24	15	.11	2	76
Alathon 7050 Solid billet	5	6.9	.3	3	80
UHMWPE Coextruded low pressure cryst.	5	.71	.07	32	60
UHMWPE Coextruded high pressure cryst.	5	6.7	.2	13	78

Coextruded Products

The extrudate prepared from the billet crystallized at high temperature and pressure had tensile modulus and strength comparable with Alathon 7050 at the same draw, but the elongation at break (12%) was much larger than Alathon 7050 (3%).

The extrudate from the billet crystallized at low temperatures and pressures was completely different. This sample was transparent and had the "feel" of a rubber. Figure 8 shows that a sharply bent sample of the "rubbery" material maintains its clarity, whereas an extrudate of Alathon 7050 shows stress crazing. The bend is recoverable in the rubbery material but not in Alathon 7050.

The mechanical data for this rubbery material are a low modulus (0.71 GPa), low tensile strength (0.07 GPa) and high elongation to break (32%). A yield point was not found and the majority of the elongation was reversible (20%). These are most unusual properties for polyethylene.

The high elongation to break of the UHMWPE extrudates from melt crystallized billets is a result of the high molecular weight of the polymer. The low elongation at break of the powder extruded product is presumably due to regions of weakness between the compacted and elongated powder particles. The big differences between the coextruded products prepared under different crystallization conditions is presumably due to morphological differences. These are under investigation.

Section 4.   The Use of Ammonia as a Plasticizer for Nylon 6 in Solid State Extrusion

The nylon family has a wide industrial application and plasticizers have been used in their processing (39,40). Even so, the prominence of hydrogen bonding between amide groups of adjacent chains makes their processing feasible only at relatively low molecular weight and high temperatures. Consequently, the choice of plasticizer depends, among other variables, on ability to withstand high processing temperature and to interrupt the interchain hydrogen bonding. Compounds effective in nylons are thus generally polar and of low volatility. A disadvantage of using such additives is that the plasticizer remains in the nylon thus tending to depress the mechanical properties. Therefore, it would be generally desirable to remove the plasticizer after processing. Volatile plasticizers can aid in processing yet be subsequently removed for enhanced final properties, however they are not readily handled in conventional drawing process.

Anhydrous ammonia has been previously used for the plasticization of wood (41,42). It further appeared to be a suitable plasticizer for the solid state extrusion of nylons. This is because it may be readily retained in the sample under pressure during

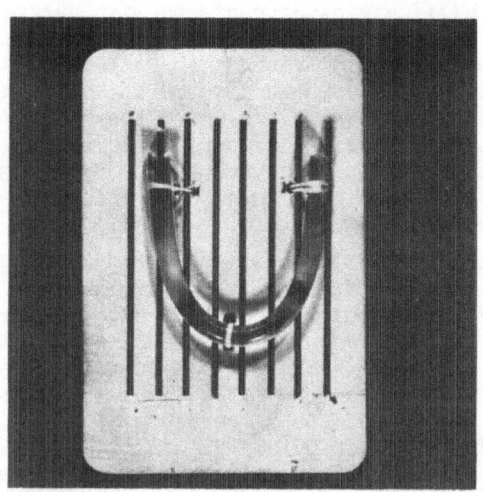

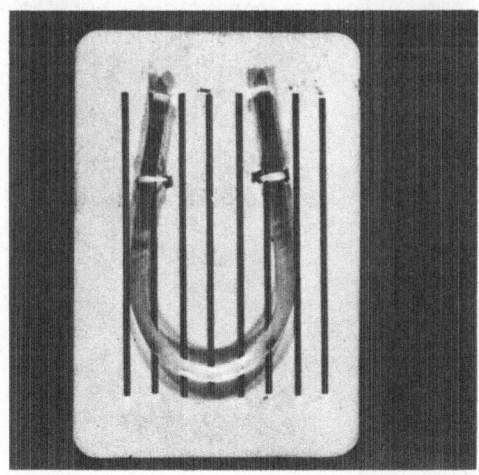

<u>Figure 8</u>: The flexibility of the "rubbery" UHMWPE prepared by
by low pressure crystallization and coextruded (top) is compared
to that of conventionally extruded Alathon 7050 (bottom).

extrusion and removed after the deformation process has occurred.
Under the conditions of our studies its incorporation in preformed nylon fibers prior to extrusion alleviates significantly
the processing difficulties encountered with untreated nylons and
provides rapid extrusion of highly oriented extrudates.   According to our technique, nylon ribbons (M_w = 16,000) 0.18 mm thick
and 9.5 mm wide were exposed to ammonia at room temperature for
5-6 hours in a pressure vessel in which the relative vapor pressure of ammonia was 1 MPa.   Thermogravimetric analysis showed
that the amount of ammonia absorbed by the nylon 6 was ~18 wt %
(see Figure 9) and that the rate of desorption, after pressure removal, increased with temperature.

The originally translucent films were treated with ammonia
and inserted within longitudinally-split high density polyethylene billets (M_w = 59,900) which were previously chilled in liquid
nitrogen.   The assembly was loaded in an Instron rheometer and extruded through a conical brass die of included entrance angle of
20° and nominal extrusion draw ratio (EDR) 12.   The extrusion
pressure was 0.20 GPa and the temperature 95°C i.e. 124°C below
the melting point of nylon 6.   On exiting the extrusion die, the
ammonia volatilized quantitatively from the deformed nylon 6.   For
comparison, experiments were conducted with untreated nylon 6
films under the same extrusion conditions, but the attempts were
unsuccessful in that no nylon extrusion occurred at the same or
higher pressures.

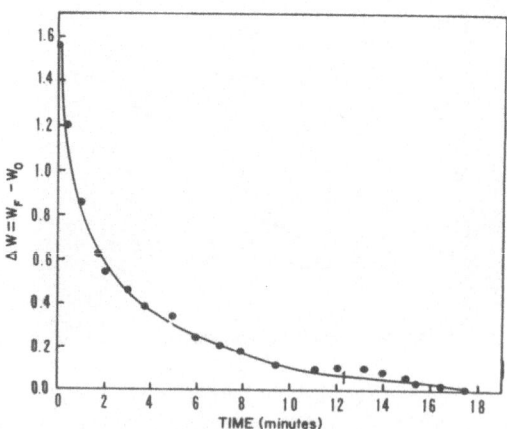

Figure 9:   Desorption of ammonia as a function of time at 50°C,
expressed by the weight loss (mg) of the nylon sample. W_F is the
weight of sample during desorption and W_o the weight after desorption.

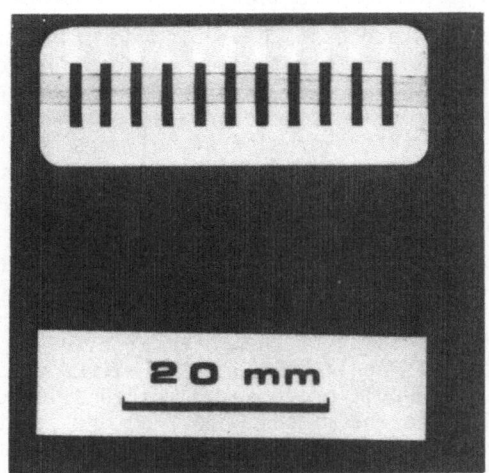

<u>Figure 10</u>:   Photograph showing the transparency of the oriented
nylon films.

In summary the experimental procedure thus involves:

1. Absorption of anhydrous ammonia at its vapor pressure
    (~1 MPa) at ambient temperature (43).

2. Retaining of ammonia in the sample by

    a. Cooling to low temperature during transfer to the
        extruder.
    b. By pressurizing at a pressure > ~5.5 MPa in the In-
        stron which is approximately the vapor pressure of
        ammonia at 95°C during the extrusion process.

3. Desorption of ammonia at atmospheric pressure at the
    exit of the extrusion die.

The oriented and transparent films obtained in these experiments shown in Figure 10 were of EDR 11.6 and preliminary property
tests indicate by thermal analysis that the crystallinity increased from 23.5% for the originally isotropic film to 53% assuming that the heat of fusion for perfect crystals of nylon 6 is
46.2 cal/g (44) and the melting point from 219°C to 223°C. The
extruded films have a tensile modulus of 13 GPa at 0.1% strain
and birefringence 8.25 x 10^{-2}. These values are higher than
those obtained previously from untreated higher nylon analogues
deformed in the solid state (45,46). The results thus document
the effectiveness of plasticization. A more detailed evaluation
of this novel plasticization process which is amenable to other
deformation techniques and to thermoplastics with interchain hydrogen bonding is under investigation.

Summary

A variety of new processes have been described that increase the range of the solid state extrusion process for low molecular weight polyethylene and also allow extrusion of a wide range of polymers, polymer morphologies and treated polymers. The factors affecting the solid-state extrusion of polymers may be divided into two main groups, factors associated with the extrusion process and those associated with the polymer itself. Similarly the processes described here may also be divided into these two groups.

The limits to extrusion of polymers by the conventional extrusion process are primarily processing factors. The extrusion pressure cannot exceed a certain value because stick-slip occurs, so eventually the viscosity of the material rises so high that the extrusion rate becomes negligible and the extrusion stops. The split billet and push-pull processes combined allowed sufficient loads to be applied to the material to fracture it at the base of the die in a form of elongational failure, i.e. the maximum draw ratio prior to fracture may be obtained. The coextrusion techniques also avoid problems associated with the extrusion process by controlling the dimensions of the components and hence limiting the extrusion pressure, and also removing shear at the edge of the films and at the surface of the core.

The other restriction on the extrusion of polymers is the mechanical properties of the polymer, a restriction that applies to all solid state deformation processes. Two developments reported here are related to the property of the polymers. By using a billet of a compressed powder rather than a melt-crystallized billet, the degree of molecular entanglement was greatly reduced and so the extrusion pressure required to deform the material was reduced. Nylon 6 could be deformed much more easily after impregnation with ammonia. The ammonia acts as a reversible plasticizer substantially reducing the force required to deform the material and increasing the elongation before fracture.

REFERENCES

1. P.B. Bowden and R.J. Young, J. Mater. Sci., 9, 2034 (1974).
2. D.M. Bigg, Polym. Eng. & Sci., 16, 725 (1976).
3. J.H. Southern and R.S. Porter, J. Appl. Polym. Sci., 14, 2305 (1970).
4. N.E. Weeks and R.S. Porter, J. Polym. Sci., A2, 12, 635 (1974).
5. N.J. Capiati, S. Kojima, W.G. Perkins and R.S. Porter, J. Mater. Sci., 13, 334 (1977).
6. W.T. Mead and R.S. Porter, J. Polym. Sci., C, in press.
7. S. Kojima and R.S. Porter, J. Polym. Sci., A-2, 16, 1729 (1978).
8. W.G. Perkins, N.J. Capiati and R.S. Porter, Polym. Eng. & Sci., 16, (1976).

9. A.E. Zachariades, W.T. Mead and R.S. Porter, "Ultrahigh Modulus Polymers", Ed. I.M. Ward, Applied Science, London, in press.
10. S. Murayama, K. Imada and M. Takayanagi, Internl. J. Polym. Mater., 2, 125 (1973).
11. A.E. Zachariades, T. Shimada, M.P.C. Watts and R.S. Porter, in preparation.
12. S. Middleman, "Fundamentals of Polymer Processing", McGraw-Hill, N.Y., 1977.
13. T. Kanamoto, A.E. Zachariades and R.S. Porter, Polym. J., accepted.
14. W.T. Mead, C.R. Desper and R.S. Porter, J. Appl. Polym. Sci. Symposia, in press.
15. C.J. Farrell and A. Keller, J. Mater. Sci., 12, 966 (1977).
16. A.S. Gibson, G.R. Davies and I.M. Ward, Polymer, 19, 683 (1978).
17. A. Peterlin, J. Mater. Sci., 6, 490 (1971).
18. E.S. Clark and L.S. Scott, Polym. Eng. & Sci., 14, 682 (1974).
19. R.S. Porter, ACS Polym. Preprints, 122 (1971).
20. L.R.G. Treloar, Polymer, 1, 95 (1960).
21. A.E. Zachariades, P.D. Griswold and R.S. Porter, Polym. Eng. & Sci., 18, 861 (1978).
22. W.G. Perkins and R.S. Porter, J. Mater. Sci., 12, 2355 (1977).
23. S.M. Aharoni and J.P. Sibilia, ACS Polym. Preprints, 191, 350 (1978).
24. A.G. Kolbeck and D.R. Uhlmann, J. Polym. Sci., A2, 15, 27 (1977).
25. K. Imada, T. Yamamoto, K.S. Migenatsu and M. Takayanagi, J. Mater. Sci., 6, 537 (1971).
26. M.A. Rudner, "Fluorocarbons", Reinhold, New York (1958).
27. E.R. Baumgaertne, U.S. Patent 3,847,888 (1974).
28. H.K. Palmer and R.C. Rowe, Powder Tech., 10, 225 (1974).
29. T. Kanamoto and R.S. Porter, submitted, Polymer Journal (Japan).
30. M.P.C. Watts, A.E. Zachariades and R.S. Porter, "New Methods of Solid State Extrusion Applied to Ultrahigh Molecular Weight Polyethylene - Processing and Properties", in preparation.
31. T. Shimada, A.E. Zachariades, W.T. Mead and R.S. Porter, J. Crystal Growth, accepted.
32. W.T. Mead, T. Shimada, A.E. Zachariades and R.S. Porter, Macromolecules, in press.
33. A.E. Zachariades, E.S. Sherman and R.S. Porter, J. Polym. Sci., B, accepted.
34. F. van Anterpen, Ph.D. Thesis, Univ. of Technology (1971), Delft, Netherlands.
35. N.J. Capiati and R.S. Porter, J. Mater. Sci., 10, 1671 (1975).
36. A.E. Zachariades, R. Ball and R.S. Porter, J. Mater. Sci., 13, 2671 (1978).

37. T. Kanamoto, A.E. Zachariades and R.S. Porter, J. Polym. Sci., submitted.

38. J.M. Lupton and J.W. Regester, J. Appl. Polym. Sci., 18, 2451 (1974).

39. M.I. Kohan, Ed., "Nylon Plastics", John Wiley, New York, 1973.

40. Encyclopedia of Polymer Science and Technology, John Wiley, New York, Vol. 10, p. 229.

41. C. Schuerch, private communication.

42. C. Schuerch, Forest Products J., 14, 377 (1964).

43. Handbook of Chemistry and Physics, CRC (1969).

44. R. Greco and L. Nicolais, Polymer, 17, 1049 (1976).

45. W.G. Perkins, Ph.D. Thesis, University of Massachusetts, 1978.

46. Y. Sakuma and L. Rebenfeld, J. Appl. Polym. Sci., 10, 637 (1966).

47. F.C. Frank, Proc. Roy. Soc. (London), A319, 127 (1970).

48. I. Sakurada, T. Ito and K. Nakamura, J. Polym. Sci., C15, 75 (1966).

49. A. Peterlin, Polym. Eng. α Sci., 9, 172 (1969).

50. S.T. Davis, R.K. Elby and J.P. Colson, J. Appl. Phys., 41, 4316 (1970).

51. A. Peterlin, J. Polym. Sci., C,9,61 (1965).

52. C.R. Desper, J.H. Southern, R.D. Ulrich and R.S. Porter, J. Appl. Phys., 41, 4284 (1970).

53. W.N. Taylor and E.S. Clark, ACS Polym. Preprints, 18, 332 (1977).

54. G. Capaccio and I.M. Ward, Polymer, 15, 233 (1974).

55. N.J. Capiati and R.S. Porter, J. Polym. Sci., 13, 1177 (1975).

56. P.J. Barham and A. Keller, J. Mater. Sci., 11, 27 (1976).

57. A.E. Zachariades, T. Kanamoto and R.S. Porter, to be published.

58. J.H. Perry, "Chemical and Engineering Handbook", 4th. Edition, McGraw-Hill, New York, 1969, Section 23, p. 31.

59. N.J. Parrat, "Fiber Reinforced Materials Technology", Van Nostrand, New York, 1972, p.68.

60. A. Zwijnenburg and A.J. Pennings, J. Polym. Sci. B, 14, 347 (1976).

GEL ENTRAPPED MULTISTEP SYSTEMS

K.F. O'Driscoll and D.G. Mercer

Department of Chemical Engineering
University of Waterloo
Waterloo, Ontario, Canada

INTRODUCTION

The immobilization of enzymes has come to be a widely practised technique (1,2). The methods of immobilization include covalent binding of the enzyme to a macroscopic substrate such as a glass or a polymer, physical adsorbtion of the enzyme on clay, glass or polymers, and entrapment of the enzyme within a polymeric network. The latter technique is of particular interest to polymer scientists because it presents us with a system where a reactive polymer (the enzyme) is catalysing the reaction of one or more substrates within the microscopic environment formed by the immobilizing network. The degrees of control over the reaction offered by the ability to vary the network properties are tantalizing.

The uses to which immobilized enzymes have been put are quite varied, but they may be regarded as falling into four categories: process catalysis, analysis, therapeutic, and scientific study of the immobilized enzyme as a model of in vivo behavior. Literally thousands of publications have now appeared describing immobilization techniques, studies of immobilized enzymes, and applications for them. It is now clear that the advantages that immobilized enzymes offer relative to the free enzymes make them an important part of the world of chemical research and industry. These advantages include the ability to reuse the often expensive enzyme and to design both the immobilization phase and the reactor so as to optimize the performance of

the enzyme. In some instances, the immobilization imparts
stabilization to the enzyme which may permit its use over
time periods of months or even years. In other instances,
immobilization serves as a means of protecting the enzyme
from a hostile environment; this is particularly true for
gel entrapped enzymes.

Given the many uses and advantages of immobilized
enzymes, it is surprising to note that, with few, notable
exceptions, only one enzyme at a time has been immobilized.
Most of the reactions catalyzed by immobilized enzyme pre-
parations have been hydrolyses or oxidations. Very few
have been synthetic reactions wherein a molecule is built
up, and even less have been sequences of reactions catalyzed
by two or more enzymes that are co-immobilized. In naturally occurring systems enzymes usually participate in
sequences or cycles of reactions. The conversion of ATP to
ADP or of NAD to NADH is often the source of free energy
which makes the overall free energy of a process favorable
in a cyclic or sequential reactions. If the full potential
of immobilized enzymes is to be realized, then we must learn
more about their behavior in multistep or cyclic processes.
In doing so we may also expect to learn more about the behavior of enzymes "immobilized" in their natural environment: the cell.

It is the purpose of this paper to review the existing state of knowledge on gel entrapped multistep enzyme
systems with particular reference to their kinetic behavior,
and to add some recent results from this laboratory.

TECHNIQUES OF GEL ENTRAPMENT

The usual method for entrapping enzymes involves preparing a solution of water soluble monomer, initiator and
enzyme in a buffer and carrying out polymerization to a
very high conversion in a short time. Initiators used are
either redox systems or photoinitiators so that the polymerization may be done at low temperature to avoid thermal
denaturation of the enzyme. Monomers used are acrylamide,
hydroxyethyl methacrylate, and other water soluble analogs.
Crosslinking monomers such as methylene bisacrylamide or
ethylene glycol dimethacrylate are usually used in 1 to 5%
amount to control network strength. Experience has shown

that the monomer is usually the most toxic reagent in the
entrapping recipes, and therefore care is taken to reduce
the exposure of the enzyme to the monomer to as short a
time as possible. Properly done, gel entrapment usually
results in the retention of 10 to 70% of the initial enzyme
activity.

The above technique leads to a bulk polymer gel swollen
with water. The gel may be chopped into fine pieces while
swollen or, alternately, dried, crushed and sieved to give
moderately uniform particles. The polymerization may also
be done in film form, thus yielding a membrane containing
the entrapped enzyme. Another popular technique involves
suspension polymerization using a non-aqueous medium such
as toluene or chloroform. Suspension polymerization is
particularly attractive because it provides an easy way to
remove heat rapidly from the polymerization and yields a
product of enzyme entrapped in spherical beads whose size
is controlled by the agitation and other conditions of the
suspension polymerization.

Other techniques which produce gel entrapped enzymes
involve dissolving the polymer and the enzyme in a common
solution and then crosslinking the polymer through some
functionality. An example of this is the use of swollen
collagen which is crosslinked with glutaraldehyde thus
entrapping a previously imbibed enzyme.

Details and specific formulations of these and other
entrapment techniques can be found in Reference 2. In
principle, there is no reason why these techniques need to
be restricted to one enzyme systems, and, indeed, they
have been extended to two or three enzymes in a number of
instances.

MULTISTEP IMMOBILIZED ENZYME STUDIES

The results of one of the first multistep enzyme systems to be immobilized were presented in 1970 by Mosbach
and Mattiasson (3). They covalently bonded hexokinase (HK)
and glucose-6-phosphate dehydrogenase (G-6-PDH) to individual polymer particles via the cyanogen bromide reaction.
Using a solution containing glucose, ATP and $NADP^+$, they
demonstrated that the coimmobilized enzymes formed product

more rapidly than an equal amount of free enzymes in a
batch reaction. However, after a lag period, the rate of
production of the product by the free enzymes approached
that of the coimmobilized enzymes. Extension of this work
to a three enzyme system where the glucose was provided
by the hydrolysis of lactose by β-galctosidase was also
achieved (4), with similar results. The conclusion was
that a steady state rate was reached more rapidly by coimmobilized enzymes simply because their proximity on the
surface produced a higher concentration of intermediates
in the boundary layer than is found in free solution. This
was in accord with theoretical predictions for surface
immobilized two step systems (5).

     Another, somewhat different three enzyme system was
coimmobilized by the Mosbach group (6). The reaction
sequence was

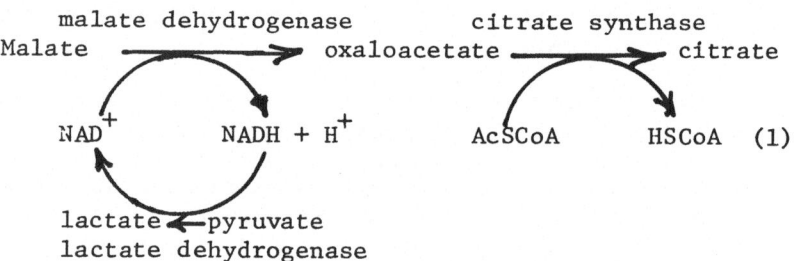

The three enzymes were coimmobilized by covalent bonding to
different matrices and also by entrapment in a polyacrylamide
gel. The latter produced the extraordinary result of a
higher steady state rate than either the surface immobilized
or free enzyme systems. The reason for this is the necessity of an intermediate product of a reactant formed inside
the gel to diffuse through the gel, past other enzymes,
before reaching the surface. The probability of it reacting
before reaching the surface is great. Further, the thermodynamically unfavored production of oxaloacetate is enhanced
by the speed with which it may be reacted in the gel entrapped system.

     Inman and Hornby (7) have reported the development of
several immobilized, linked enzyme systems for use in the
automated determination of disaccharides. The appropriate
enzymes are bound to the inner walls of nylon tubing which

is chemically activated. One such system incorporated
invertase and glucose oxidase to detect sucrose via the
resultant depletion of dissolved oxygen in the sample
stream flowing through the tubing.

$$\text{Sucrose} \xrightarrow{\text{Invertase}} \alpha\text{-D glucose} \rightleftharpoons \beta\text{-D glucose}$$

(2)

$$\beta\text{-D glucose} + O_2 \xrightarrow{\text{Glucose Oxidase}} \text{Gluconic acid} + H_2O_2$$

Suzuki and coworkers extended this work to produce an enzyme electrode for the measurement of sucrose (8). They
incorporated invertase, glucose oxidase and a mutarotase
(to catalyze the α-β equilibrium) in a collagen membrane
stretched over the oxygen permeable Teflon membrane of a
conventional oxygen electrode. This devise is capable of
continuously monitoring a sucrose solution concentration.
Recently, Cousineau and Chang (9) have shown that glutamate
can be synthesized in a microencapsulated, multistep enzyme
system. Urease, glutamate dehydrogenase and G-6-PDH were
coimmobilized in microcapsules to convert α-ketoglutarate
to glutamate.

In another model study, Mosbach et al (10) have shown
that a proton generating enzyme can change the relative
ability of two enzymes to compete for the same substrate.
They coimmobilized hexokinase and glucose oxidase, which
have pH optima of 8.5 and 6.6 respectively, along with
trypsin by entrapping them in a polyacrylamide gel. Reaction at a supernatent pH of 8.5 with a limited amount of
glucose present permitted the hexokinase to phosphorylate
about 13% of the glucose being consumed, the remainder
being oxidized by the glucose oxidase. When a substrate
for the trypsin was added, the hexokinase activity was
observed to fall almost to zero. Presumably, the proton
generation of the trypsin catalyzed reaction caused a micro-
environmental change in the pH, thus shifting the local pH
in the gel more toward the optimum of the glucose oxidase
and away from that of the hexokinase. This demonstrates a
possible switching or regulating mechanism in the cell.

Thomas and coworkers have also performed modelling
studies, using enzymes coentrapped in membranes (11,12).
For example, salicin used as a substrate in a membrane

containing β-glucosidase and glucose oxidase produces glu-
conalactone as an end product, which is an inhibitor for
the first enzyme. Thus there is a feedback inhibition
operating in this system which was shown to be more pronounced in the immobilized case than in free solution.
Again, the localized concentration affects the relative
performance of free and coimmobilized enzymes. In this
case, the kinetic results were successfully simulated by
taking into consideration the coupled nature of the diffu-
sional processes and the enzyme catalyzed reactions.

Theoretical studies have been published by Gondo (13)
and by Ramachandran and coworkers (14,15,16). Gondo showed
that the effects of diffusion on multiple immobilized
enzyme systems were such that mass transfer resistances are
expected to improve the yield of final product. Ramachandran has applied the interior collocation technique of
Villadsen and Stewart (17) to account for the expected
influences of intraparticle and interparticle diffusion on
the kinetics of immobilized two step enzyme systems.

In all of the multistep immobilized enzyme work done
to date, theoretical or experimental, for modelling purposes or for applications, there exists one common factor:
the chemical reactions are affected by the diffusive
processes so that the macroscopically observed kinetics are
strongly perturbed by the incorporation of the enzymes into
a gel. This perturbation is caused by the development of
localized concentrations and concentration gradients within
the gel which are quite different from that found in free
solution. Only one instance appears to have been reported
where exact modelling of real experimental data has been
attempted. All other work has been either purely theoretical or qualitative interpretations of limited experimental
data. There is still much to be learned of the role played
by the gel matrix in affecting the overall kinetic performance of gel entrapped multienzyme systems before they can
be well designed for applications or used with any confi-
dence in a quantitative way as models for living systems.

What is needed then is a series of studies where the
observed kinetics are explained by exact modelling which
quantitatively accounts for not only the diffusive behavior
but also the gel structure and its influence on the chemical
reactivity of the substrates, the reactants and the enzymes.

As a start on such a series and by modifying the kinetic treatment of Ramachandran while retaining the use of the collocation method, the present authors have successfully analyzed experimental observations made on the system of reaction scheme (2). Both invertase and glucose oxidase were coimmobilized by entrapment in gels of poly(hydroxy ethylmethacrylate) (PHEMA) or of poly(hydroxy ethylmethacrylate-co-4-vinyl pyridine) (PHEMA-VP). Some of the more significant results are presented below; for details and more extensive results, the reader is referred elsewhere (18,19,20).

INVERTASE AND GLUCOSE OXIDASE COENTRAPPED IN PHEMA

Summary of Experimental Work

A detailed kinetic investigation was made on gels of PHEMA and PHEMA-VP containing 2.5% VP. The gels per gram contained approximately 4 units of invertase and 9 units of glucose oxidase activity and had approximately 30 and 40% water respectively. The vinyl pyridine was put into the gel in an attempt to enhance the mutarotation of the α-D-glucose to the β form. Experiments were performed in a small tubular reactor into which was pumped an air saturated solution of sucrose of known concentration. The effluent oxygen concentration revealed the extent of β-D-glucose oxidation, while a glucose oxidase containing gel in another reactor could be used to monitor the effluent β-D-glucose concentration. The same glucose oxidase reactor could also be used to measure the total α- and β-D-glucose concentration of the effluent by allowing the effluent to equilibrate for a time sufficient for the α-β equilibrium to be attained. In this manner, all concentrations of interest (sucrose, α-, β-D-glucose and products) could be computed from a given set of experimental data. A typical set of results are shown in Figure 1. The solid lines in Figure 1 are theoretically calculated as described below.

In order to model the behavior of the reactor, the enzyme activities and the mutarotation rate constants had to be separately determined. The invertase activity was determined by measurements similar to those described above; the glucose oxidase activity was determined in the

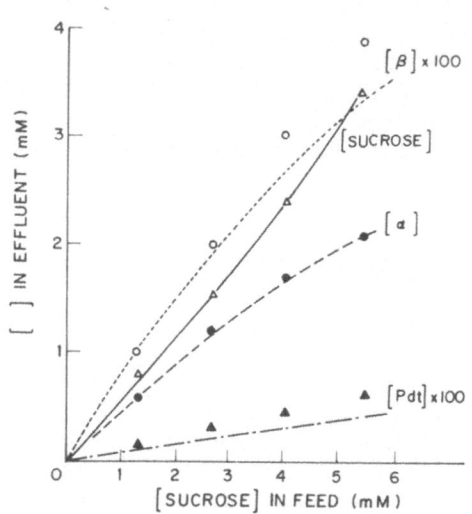

Figure 1. Concentrations in effluent vs Sucrose feed
concentrations for PHEMA gel, 149-212 μm particles in 12 cm
reactor, 0.4 cm I.D. at flow rate of 0.7 ml/min. Solid
lines are result of collocation analysis.

same gel by feeding it a solution of known glucose concentration. The mutarotation rate constants were determined
by polarimetry in an experiment described in detail elsewhere (19). Polarimetry results, shown in Figure 2,
clearly indicated that the gel containing VP allowed
mutarotation to occur more rapidly than did the PHEMA gel.
When quantitative allowance was made for the mutarotation
occurring in the supernatent buffer and for the difference
in water content of the gel, the startling result was
obtained that no mutarotation occurred in the gel phase of
PHEMA! In contrast, the PHEMA-VP gel had mutarotation
rate constants equal to those in the buffer. The equilibrium proportions of α- and β-D-glucose were unaffected by
the presence of either gel.

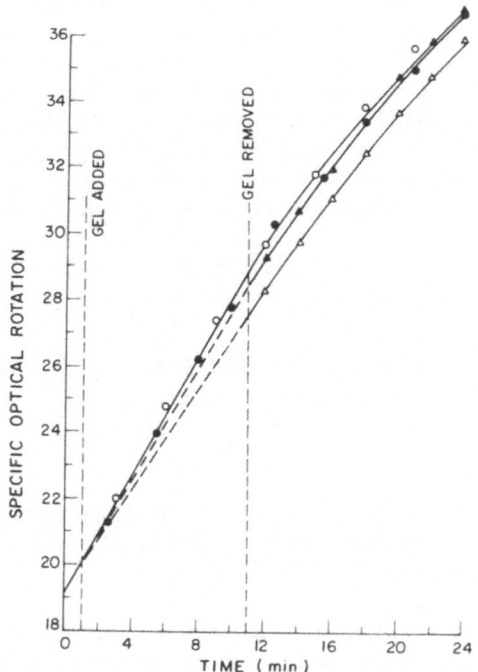

Figure 2.   Optical rotation of β-D-glucose as a function of time in buffer (○,●) and in supernatant of slurries where gel was present from 1 minute to 11 minutes: (▲) PHEMA, (△) PHEMA-VP.

Summary of Kinetic Treatment

The gel particles containing the two enzymes were assumed to be spherical, and the concentration profile of each species of reactant and product was calculated within a bead using the collocation technique (17). In a given bead, any species i is both diffusing and reacting so that its concentration at any point satisfies the equation

$$\frac{D_i}{r^2} \left(d(r^2 \frac{dc_i}{dr})/dr \right) = f_i \tag{3}$$

where D_i is the diffusion constant of species i having concentration c_i at radial position r, and reacting at a rate given by f_i. It is important to note that f_i may be a function of the concentration of species other than i. The boundary conditions on equation 3 are that the concentration gradient is zero at the center of the sphere and that the flux of species i to the surface from the interior of the bead equals that through the boundary layer:

$$\frac{dc_i}{dr} = 0 \text{ @ } r = 0; \quad D_i \frac{dc_i}{dr} = k_b (c_{i,b} - c_{i,R}) \text{ @ } r = R \quad (4)$$

where R is the radius of the particle and k_b is the mass transfer coefficient in the boundary layer. The supernatent concentration of i is $c_{i,b}$ and that at the particle surface is $c_{i,R}$.

Equation 3 can be solved by the collocation technique in the following manner, which is described here in a very oversimplified way:

A number of points are chosen in the spherical particle. These N collocation points are such that they define shells of equal volume. N is typically 3 or 4 and includes the surface. For i species there will then be $N(i-1)$ equations of the form

$$a_i c_i + b_i - \frac{R^2 f_i}{D_i c_{i,b}} = 0 \quad (5)$$

where a_i and b_i are numbers whose values may be calculated from tabulated parameters for each collocation point (17) and from the Sherwood number, $Sh = k_b R/D_i$. Solutions of the simultaneous equations (5) yields values of the concentrations c_i at each collocation point. A typical result is presented in Figure 3.

Since there is a void volume associated with each particle, the flux from the first particle in the tubular reactor determines the concentration in the supernatant which is the feed to the second particle in the reactor. In this manner, and allowing for some feedback with a "tortuosity factor" of 2, the concentration of each species

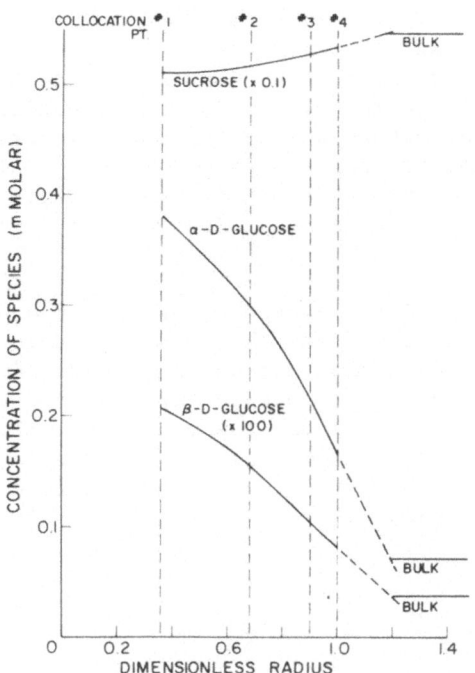

Figure 3. Collocation solution of a particle having 115 μm
radius. This particle was the 23rd from the entrance of a
tubular reactor to which 5.5mM sucrose was fed.

can be computed, bead by bead, along the entire length of
the reactor. The effluent concentration calculated in this
manner is the solid line in Figure 1.

Summary of Results

Numerous experiments were performed, and the results
compared with the model. Parameters varied included VP
content of the gel, particle size, flow rate, reactor
length. In every case sucrose feed was varied and the
effluent concentrations compared with the theoretical
predictions. The model usually gave results with a precision of 5% or better for all species, under all conditions measured (18,20). It is important to note that the

results with the PHEMA gels were properly described with
the assumption that <u>all</u> mutarotation occurred in the
supernatent liquid, so that a given α-D glucose molecule,
formed by hydrolysis in one particle had to diffuse out
of that particle, mutarotate in the supernatnet, and then
diffuse into another particle before being oxidized to
product. For the PHEMA-VP gels it was possible for mutarotation and oxidation to occur in the same particle as
hydrolysis.

Given the excellent agreement between the experiments
and the model, the model was then used to perform
"experiments" on the computer; parameters could be varied
in the model in ways that were interesting for the information they gave, even though the actual experiments would
have been difficult or impossible. For example, Figure 4
shows the effect of varying the ratio of enzyme activities
on the yield of product.

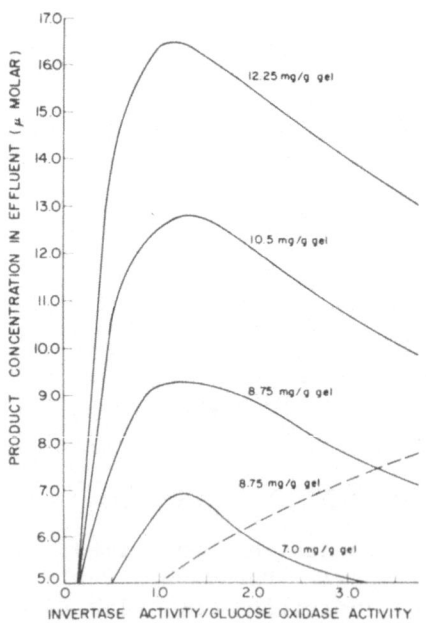

Figure 4. Calculated effect of enzyme activity ratio on
product yield at fixed experimental conditions. - - - -
PHEMA gel; ----- PHEMA-VP gel

For the PHEMA-VP gel the existence of a maximum is observed, which is readily understood as the result of a transition from invertase to glucose oxidase being the rate controlling reagent. In PHEMA, however, no maximum is observed and this is because there is so little mutarotation occurring, all outside the gel phase, that it is always the rate controlling step.

Another interesting result from model simulation involves the importance of the diffusivity possible in the gel phase. In practice the diffusivity in the gel is mostly a function of water content and crosslink density. However, any experiment designed to measure the variation of reaction in the gel as a function of diffusivity would be confounded by the occurrence of numerous other changes. Figure 5 shows the calculated effect of diffusivity in two gels, curve A representing PHEMA-VP and curve B, PHEMA.

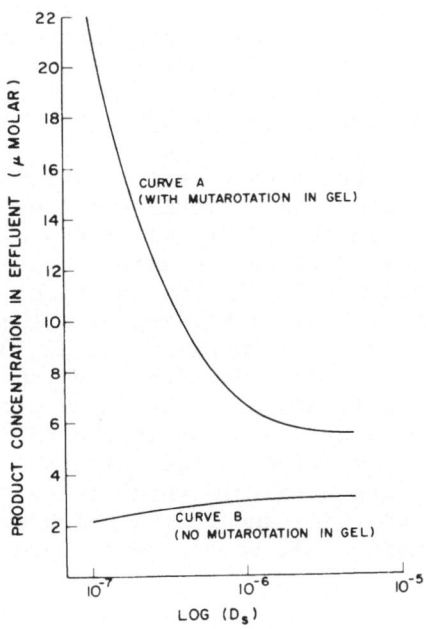

Figure 5. Calculated effect of diffusivity in gel on product yield at fixed experimental conditions.

In the latter, the effect of diffusivity on the observed product formation is almost nil. This is again interpreted as being due to the extra-particle, rate limiting step of mutarotation. In Curve A, the dependence on diffusivity is quite strong and in the direction opposite to that suggested by Gondo (13) as a generalization. This particular reversal of Gondo's generalization occurs because restricting the mobility of the α-D-glucose permits it more time to mutarotate within the gel particle where it was formed and thus increases the concentration of the β-D-glucose within the particle. This in turn enhances its rate of oxidation to product.

CONCLUSIONS

The quantitative data and observations discussed above appear to be consistent with the qualitative concepts put forward by Mosbach and others to explain the behavior of multistep enzyme systems. Mathematical modelling of gel entrapped, multistep, immobilized enzyme systems, using the collocation technique, is quite straightforward and provides the opportunity not only to compare experiment with theory but also to explore the effect of parameters of the system which are experimentally inaccessible.

A surprising, but important fact which emerged from the PHEMA study described above is the lack of chemical activity of water in the PHEMA gel in the mutarotation reaction of glucose. Since the water is a chemical reagent and not just a solvent in the mutarotation, it is reasonable to assume that the physical binding of water to the gel as described by Andrade and coworkers (21,22,23) affects the chemical reactivity. The addition of VP to the gel appears to reverse the deactivation of the water within the gel, causing the rate of mutarotation in the PHEMA-VP gel to approach the rate in buffer. This might be attributed to a combination of the catalysis which is to be expected of the pyridine moiety (24) and to the enhanced water content of the VP containing gel, some of which presumably is not bound water.

The effect of the gel on the reactions within it is a clear example of a "microenvironment". Other effects which polymers have been suggested to exert on immobilized enzyme activity include excluded volume (25,26), conformation

alternation (27) and microcompartmentation (28). It is
possible that changes of activity of water in immobilized
enzyme systems or even in whole cells, caused by the
presence of water binding polymers, may play an important
role in the enzyme activity of such systems.

ACKNOWLEDGEMENT

Support of this work by the Natural Sciences and Engineering Research Council of Canada is gratefully acknowledged.

REFERENCES

1. O.R. Zaborsky, Immobilized Enzymes, C.R.C. Press,
    Cleveland, Ohio, 1973.

2. Methods in Enzymology, v. 44, Immobilized Enzymes,
    K. Mosbach (ed.), Academic Press, N.Y., 1976.

3. K. Mosbach and B. Mattiasson, Acta. Chem. Scand., 24,
    2093 (1970).

4. B. Mattiasson and K. Mosbach, Biochim. Biophys. Acta.,
    235, 253 (1971).

5. R. Goldman and E. Katchalski, J. Theor. Biol. 32, 243
    (1971).

6. P.A. Srere, B. Mattiasson, and K. Mosbach, Proc. Nat.
    Acad. Sci. U.S., 70, 2534 (1973).

7. D.J. Inman and W.E. Hornby, Biochem. J., 137, 25 (1974).

8. I. Satoh, I. Karube, and S. Suzuki, Biotechnol. Bioeng.,
    18, 269-272 (1976).

9. J. Cousineau and T.M. Chang, Biochem. Biophys. Res.
    Commun., 79 (1), 24-31 (1977).

10. S. Gestrelius, B. Mattiasson and K. Mosbach, Eur. J.
    Biochem. 36,89 (1973).

11. J. Hervagault, M.C. Duban, J.P. Kernevez and D. Thomas,
    J. Solid Phase Biochem. 1 81 (1976).

12. J.F. Hervagault, G. Joly, and D. Thomas, Eur. J.
    Biochem. 51, 19 (1975).

13. S. Gondo, Chem. Engng. J. (Lausanne), 13, 153–163
     (1977).

14. R. Krishna and P.A. Ramachandran, J. Appl. Chem.
     Biotechnol., 25, (623–640) (1975).

15. P.A. Ramachandran, R. Krishna, and C.B. Panchal,
     J. Applied Chem. Biotechnol, 26, 214–224 (1976).

16. P.A. Ramachandran, Biotechnol. Bioeng., 17, 211–226
     (1975).

17. J.V. Villadsen and W.E. Stewart, Chemical Engineering
     Science, 22, 1483–1501 (1967).

18. D.G. Mercer, Ph.D. Thesis, University of Waterloo,
     1978.

19. D.G. Mercer and K.F. O'Driscoll, J. Molec. Catal. 4,
     217 (1978).

20. D.G. Mercer and K.F. O'Driscoll, Biotech. Bioeng.
     (in press).

21. M.S. Jhon and J.D. Andrade, J. Biomed. Mater. Res., 7,
     509–522 (1973).

22. H.B. Lee, M.S. Jhon, and J.D. Andrade, J. Colloid and
     Interface Sci., 51 (2), 225–231 (1975).

23. H.B. Lee, J.D. Andrade, and M.S. Jhon, Polymer Pre-
     prints, 15 (1), 706–711 (1974).

24. A.S. Hill and R.S. Shallenberger, Carbohydrate
     Research, 11, 541–545 (1969).

25. T.C. Laurent, Eur. J. Biochem., 21, 498 (1971).

26. B. Mattiasson, A–C. Johansson and K. Mosbach, Eur. J.
     Biochem., 46, 341 (1974).

27. V. Jancsik, T. Keleti, Gy. Biczok, M. Vagy, Z. Szabo
     and E. Wolfram, J. Mol. Catal., 1, 137 (1975).

28. P.A. Srere and K. Mosbach, Ann. Rev. Microbiol. 28,
     61 (1974).